On the Verge of Tomorrow Renewable Energy: An Introduction for Families and Communities

Marc D. Rappaport

Disclaimer

Renewable Energy: An Introduction for Families and Communities

Printed in the United States of America.

First Printing: June 2017.

Dedication

To my parents, Samuel, and Ruth Rappoport;(Solomon)
My uncle, Morton Leeds, PhD;
My grandchildren Lucy and Theo.

Contents

Preface

One of my favorite things to do when I was young was making things, whether my erector set, models of planes, boats, cars—anything. As I got older, I started going to drawing and drafting classes. In junior high school, I had some of my designs exhibited for homes, and so, designing and building have been in my background since my learning age.

This book is going to include the discussions of my development of one of the very first, fully independent biomass cogeneration power projects. It utilizes woody waste (biomass) debris from community sources to power the power plant, and it is still functioning in Medford, Oregon even after 30 years. It takes 280,000 tons out of the waste stream and runs it through the power plant with state-of-the-art emission controls. The community itself delivers 80,000 tons of woody waste (biomass) debris and it replaces burning in the backyard or delivering this to landfill to make the local community power.

I think this is an excellent format to demonstrate the viability of community-scale distributed energy with decentralized power. This provides the local community a source of extra power in the event of a system-wide failure, which there was during a winter storm when lightning struck high-voltage power lines that bring power into the Medford Valley.

The energy industry in this country is enormous. Engineers, scientists, and manufacturers are all involved to keep our national grid up and going—the value of it is always understood when we go to the refrigerator, get a cold beer, or when we turn on the lights at night.

This book is an effort to distill down the fact that renewable energy can be a promise for our country and the planet. I am of the opinion that it is enormously valuable for local communities. This book is part of what the whole effort in renewable energy is about. In it, I will explain my thoughts and successes, illustrate examples that work for communities and individuals at large, and explain how it has helped reduce emissions in many areas.

Acknowledgements

I have been interested in solar energy for many years. For 50 years, I have met and talked with hundreds of people at fairs, conferences, and various meetings and training. It would be impossible for me to name them, but it is with a profound sense of appreciation that I thank them for discussing their work and mine. From a time when there was just a small group organizing the Oregon Solar Energy Association in the 1970s to today's conferences on renewable energy where there are thousands, it is great to see. I remember one time testifying at a state energy committee when an older legislator was questioning me about the possible future of solar energy and asking sarcastically if I thought we would drive cars with solar energy. I answered *yes*. I wonder what he would have said if I had answered that in the future where people would fly around the world with a solar-powered airplane! The *Solar Impulse* airplane did do that.

Introduction

This book is written as a result of years of conversations with people, doing public talks, and public testimony which helped inform people about the true value of renewable energy. This book is written for the person who cares about the environment and wants to become more informed on what he/she can do to contribute toward change and a better life.

I am writing this book is to share some of my experiences and insights from many years of working in renewables with a greater audience since not many people have heard of a biomass co-generation power plant that is not part of a wood products company or exposed to the idea of a geothermal solar thermal hybrid system for power generation as well as the multitude of choices for energy conservation and efficiency. A book seemed the best way to disseminate my ideas and experiences. Using the refrigerator list, you can take small steps for a more energy-efficient home. Taking this approach with some patience, you will recover the cost of this book many times over the years you live in your home.

Marc Rappaport: Biography

Mr. Rappaport has been involved in renewable energy for over forty years. He has had various roles in different fields with interesting and varied success. He has designed solar buildings and had been invited to present one at an AIA conference in Seattle during the 1970s. He provided architects with solar building optimization analysis and built a solar water heater from recycled material in 1971 proving you could get hot water in Oregon from solar energy. He has also served as a founding president and a charter member of the Oregon Solar Energy Association.

In 1972, he invented the geothermal-solar hybrid system for which he was awarded a basic patent in 1978. He was awarded a proof-of-concept grant by the Carter administration in 1979. However, the award was cancelled by the Reagan administration as part of the cuts to renewable energy research. During the development of the geothermal-solar system, Mr. Rappaport engineered a biomass power project and developed its business plan as well. From this work, the development of Biomass One, LP emerged, and the construction of a 22 MW plant has now evolved into a 30 MW biomass plant in Medford, Oregon that has been in commercial operation for the past 30 years. This project can supply power to 30,000 households. He has served as a founder, general partner, and a managing partner of Biomass One, LP. Now in retirement, he has been working on the development of a biomass refinery and cogeneration technology.

His work in sustainable energy includes speaking at various conferences and having peer-reviewed articles published in a number of proceedings where he was a speaker. He has served on the Western Governor Association,(WGA), Biomass Task Force, the Portland Energy Commission, Solar Task Force, chaired the ACORE (American Congress Of Renewable Energy) WIREC (Washington International Renewable Energy Congress) session on Bio-Refinery Green Fuels, and had been the only American invited to present at the Sustainability of the Planet conference in Stockholm, Sweden held at the Nobel Institute in 2007. He is a member of ACORE, NY Academy of Sciences, a past member of the Geothermal Resources Council (GRC), International Society of Sustainability Professionals (ISSP), and has been included in Who's Who Worldwide in the 1994–1995 edition. During the Clinton administration, he was a participant in the biomass task force for the transition team and continued to provide input to the biomass review committee for eight years, which included a white paper for the White House Office of Science and Technology Policy (OSTP) that resulted in a Presidential Executive Order creating the Biomass Coordinating Committee.

The following is a brief outline of the hybrid system's benefits:

- Integration of a Geothermal and Solar Thermal System to Upgrade Non-Flashing Geothermal
- Most geothermal water is not hot enough for creating flashing steam to drive a steam turbine.
- A geothermal field can have multiple wells at some distance from the extraction wells, sometimes miles.
- Solar thermal systems are expensive for power generation.
- Collecting wells to a central plant is common.
- Using linear concentrating solar thermal for temperature boosting uses a fraction of a stand-alone solar-sized system for the same output when combined with geothermal.
- Site specific for each geothermal field.
- Can be started with existing KGRA data.
- Existing wells can be analyzed & prioritized.
- Formulas allow for the optimization of solar.
- Compatible with CHP (Combined Heat & Power) systems.
- Allows for reinjection into the field for long-field life & prevents pollution.
- Design allows for extended system life with upgrades and maintenance for a long amortization and payoff.
- Systems can be designed for community integration and uses.
- Can be designed & engineered for binary or tertiary systems.
- Adds distributed generation to the grid.

Chapter 1: Beginning to Understand Renewable Energy

We will be looking at communities that have already adopted successful conservation, weatherization, and renewable energy programs which can now also be transferred to your local community or state helping families and communities at large save energy, money, and thereby reduce the CO2 footprint.

In this book, you will find useful information offered in an understandable format for you to take action. Information in this book should also make you able to recover the expense of this book with the energy conservation ideas you will find herein.

There is copious research material available on this subject. However, with all of the information I have gathered over the years, I have made the effort to condense the material and have given you something that is immensely useful for your state, community, and your home.

State Programs

All state programs begin with state legislation and the public utility commission of the state. Public utility commissions regulate the utilities within the state and there are a couple of entity models; one of which is the public utility; another is called investor-owned utilities (IOUs). The IOUs are the private utilities within a state.

The biggest example of an IOU is Pacific Gas & Electric in California. It is one of several investor-owned utilities in the state that provide services to the public and is regulated by the public utility commission. If you live in a different state, you can look up the public utility commissions and find out which utilities are IOUs or public utilities. Oregon, for example, has the Oregon Public Utility Commission and regulates Portland General Electric and Pacific Power & Light which are IOUs. There are also several public utilities including Eugene Water & Electric Board. The state of Washington has Puget Sound Energy as its major IOU; it has numerous public utilities. Seattle City Light and Clark Public Utilities are examples of its public utilities.

In these three states, the Public Utility Commissions have established policies and programs that incentivize the utilities to support conservation efforts, including replacing incandescent light bulbs with LED bulbs, implementing some weatherization programs, replacing appliances with their more energy-efficient counterparts, and providing some cash bonuses for adopting these programs.

Chapter 2: Renewable Energy Technologies Are Numerous

Why do people only think about solar energy as comprised of photovoltaic panels and wind? There are many more options available. There are two types of solar energy: photovoltaic and thermal. Solar photovoltaic energy converts sunlight directly into electricity, whereas solar thermal energy harnesses the heat generated by sunlight.

Solar energy technologies include solar water heaters, pool heaters, and supplements to ground source heat pumps. There are solar-powered photovoltaic (PV) panels which can be roof-mounted, ground-mounted single-axis solar trackers, and dual axis solar trackers. Larger facilities could be decentralized for distributed generation as community-scale power systems. You also have solar thermal systems technology that can supply power and process hot water for various commercial needs. There are various designs for each of these systems in the form of concentrating solar power towers or a linear concentrator that uses parabolic trough collectors (CSPs).

Solar designs for houses and buildings can mean passive solar heating. There is a passive solar wall design called a Trombe wall where the thermal energy absorbed by the Trombe wall can be stored in the thermal mass that exists inside the building. It can be stored in eutectic salts or other phase change materials (PCMs) in rocks under the floor or in a wall. These are just a small number of solar designs available.

Along with solar power, biomass power from woody waste debris can also provide electrical system power. Biomass cannot always be seen, but it is present at sawmills, especially at pulp and paper mills where they generate their own power. Often, people have wood stoves and they are fueled by pellets or chopping wood; this is an exact example of biomass power and heat.

Another subject area that we will explore is geothermal energy. Geothermal energy can provide space heating with heat pumps, power from organic Rankine cycle power systems, and geothermal production from high-temperature power plants as well.

High-temperature power plants are found in California in the Geyser fields and in the Imperial Valley, in Iceland, and in various other countries across the globe. It was in Italy on the Larderello geothermal field where the world's first geothermal power generator was tested. It has been operating for well over 100 years, and the field is now home to a total of twenty operating geothermal power stations. In New Zealand, the first power station to use wet steam was built near the Waiariki geothermal field and was phased out in 2013. It was replaced by the Te Mihi geothermal power station. Wairakei lies in the Taupo Volcanic Zone; the power produced there from geothermal energy is very high.

I will be explaining a hybrid geothermal solar thermal power plant in detail in the following chapters as I am the inventor of the system. I was awarded the patent for it in 1978. I invented this system in May of 1972 and have seen enough events occur that I am moving forward to bring this technology to broader exposure, having successfully seen the adaptation of the hybrid system fix a failing geothermal plant in Nevada run by Enel, and to provide enough energy to bring the plant into full compliance with its power sales contract.

Another new technology that is just getting started and has proven successful is an in-pipe hydro generation—the little generators that sit atop pipes where a water system has water constantly flowing through it, spinning the blades that turn the generator that produces electricity. The City of Portland Water Bureau had one of these systems recently installed in their system bringing water from the Mt. Hood water reservoirs down to Portland.

Another energy-saving system is district heating from a central steam plant. This is a system for distributing the heat generated in a centralized location by way of insulated pipes for residential and commercial heating needs such as space heating and water heating.

Another form or example of district heating with geothermal energy is the town of Klamath Falls. There are buildings heated in Boise, Idaho, and Reno, Nevada whose energy requirements are largely supplied by geothermal as well as district heating from geothermal. One of the larger hotels and casinos in Reno, the Peppermill Resort Spa Casino, is heated entirely by geothermal energy.

A variation of the geothermal energy is a geothermal heat pump that draws heat from the ground to provide it to a house. In the summer, it does the reverse where it takes the heat out of the house and puts it into the ground as it is providing cooling.

Another source of energy is called fuel cells—some examples are solid oxide fuel cells that run on hydrogen and are being installed in automobiles; their emissions are water vapor. In further explanations, I will add more information about a half dozen fuel cells and technology from the Department of Energy for your understanding.

Another form of energy saving is the weatherization of our buildings. If your house is an older construction, you will probably benefit from weatherization strategies. Add weatherization insulation in the walls and doorways that go outside which need weather stripping. For instance, during wintertime, you would want to close off the crawl space vents. Also, some people are replacing their old aluminum single pane with thermal dual pane windows, and while that may be more expensive than you'd like, there is a company that is making a new design of an interior window called Indow, which is comparatively more affordable. This product is called the Indow window.

There are opportunities for weatherization and energy conservation in lighting and in sealing up cracks. Hopefully, there will be coupons available for the book that will make it easier for you to go down to your local hardware store, whether it is Home Depot, Loew's, or Ace Hardware, to purchase a can of insulating foam. You will be able to put this in narrow cracks where leaks exist and seal them up.

Another form to make energy available for when we need it is pumped hydro. It can be done in a period when there are excess energy and sufficient water with a place to store it. Pumps take water from the lower level and pump it up into a higher reservoir, and when the demand requires more energy, the water is released to flow through turbines that produce the electricity at the time the demand is called for.

Additionally, there is now progress in thermal storage as well as battery storage for photovoltaic systems.

Another aspect and new design are net-zero energy homes and buildings—these are buildings and homes that have been specifically designed so they are net-zero of energy demand in the community on the grid.

Engineers and inventors are continuing to make innovative technologies. There are more coming.

Chapter 3: The Utility Industry

The utility industry companies operate as monopolies, and so, the use of public utility commissions is needed to regulate their industry. Further, the states and utilities are regulated by the Federal Electric Regulatory Commission (FERC). There is layer upon layer of regulatory oversight in this industry.

Regulatory rules play an important role in the utility industry. One US act called the Public Utility Regulatory Policies Act (PURPA) of 1978 mandated that utilities would have to buy power from independent power producers or IPPs, which would be able to source their power from renewable energy.

Until this legislation passed, the utility industry was unwilling to enter into contracts that would be financeable for developers of IPPs; hence, the reason for the legislation to come into existence. Once that legislation was passed, it did not take very long for the utilities to find ways to oppose it. In doing so, the states created what is called renewable portfolio standards (RPS) to mandate that utilities buy a certain percentage of new energy from renewable energy producers.

It became a fight at the state legislatures that utilities would have to have a portion of their power come from renewable energy.

Today, what we have in California is a 50% RPS, and in Oregon, we have a 50% RPS. This means that new power going into production must have 50% of that power coming from renewable energy. This, of course, is a huge benefit for renewable green energy and has led to significant cost reductions in that industry. These cost reductions have been so significant that today, photovoltaic energy is entirely competitive with coal and natural gas or fossil fuels. Policy is changing today in various states, so there will be new standards and goals that are not identified here. California has adopted a Cap & Trade program to lower carbon emissions along with several northeastern states.

One of the things to understand is that these are not just the utilities—there are investor-owned utilities (IOUs) along with the public utility power companies operating in this field all across the country with billions upon billions of infrastructure costs involved.

One of the main centers of research for the utility industry is the Electric Power Research Institute or EPRI. In addition, you have the Rural Electric Administration (REA), the Tennessee Valley Administration (TVA), the Bonneville Power Administration (BPA), and the independent power producers (IPPs). The REA, TVA, and BPA primarily are involved in some power generation but to a large degree—in the transmission of distribution systems and the intertie (which is an interconnection that permits the passage of current between two or more electrical utility systems). One of the ways that power comes into various states is through the interstate intertie. Large towers that you have seen as you drive around have high-voltage power lines. These power lines are used to move electrical current to substations, which then distribute power to communities through the wires out to your neighborhood and to your house via transformers. I need to mention that dams on the Columbia River are operated by the Corp of Army Engineers and elsewhere.

What you need to know is that your individual choices can make a difference in how much power is used. Think about the simple step of changing a light bulb. If you change one incandescent light bulb—those that you have been using for years and years—and substitute it with a new LED light bulb, the difference is quite significant. That switch from one light bulb technology to the newest light bulb technology has been calculated to reduce emissions in coal plants by *500 pounds per year.*

One of the things I am going to try and do with this book is to establish a Google website that will have a continuous counter, so you can go to that website and enter the number of light bulbs you have purchased or gotten for free from your utility and have installed in your house. What we will be able to see is the cumulative effect of everybody's actions together and the impact on the reduction of emissions. This way, you will know that your actions are bringing about positive change in the world. Our actions may seem insignificant, but when added together, a simple step has a huge impact.

One of the other major things you can do is to add insulation simply around your water heater. You can go to Loew's, Home Depot, or Ace Hardware and get a water heater insulation kit or a water heater blanket. Simply wrap it around your water heater and that action itself will begin to save you energy. As mentioned before, there is a new manufacturer in Oregon called Indow that makes an insert for the interior of your house that you place over your windows to create a double-paned window. It is less expensive and easier to install than new thermal pane windows, and significantly benefits the environment; it also reduces noise pollution. Alternatively, if you were to use the small cans of foam insulation around big leaks in your older house, this again will slow the infiltration of cold air and save you on your heating bill.

A little-used solution to heat gain and the sources of heat if you live in a southern climate and have a flat roof is to paint your roof white. It is that simple. Another tip is that the next time you need to waterproof your roof, make sure to pick out a product that is rubberized white paint and you will solve two problems at once: heat gain and leaky roofs. Hopefully, there will be additional coupons available from the stores I've mentioned so that the book pays for itself. Take a simple action: weatherize your house and make it energy-efficient.

Chapter 4: Defining the Problem of Finding New Energy

Over the years, one of my concerns has been the lack of questioning the long-term value of nuclear energy as a universal solution to energy problems. As a high school senior student, I made a trip to the nuclear research facility at the University of California, Berkeley campus. There, I operated an experiment with the cloud chamber device that showed the pathways of subatomic particles after a single atom is hit with a neutron. Witnessing the arrays as tracer pathways was something I have never forgotten. However, this was not my first introduction to nuclear research.

When I was about seven years of age, my father took me on a trip to Princeton, New Jersey where we met a friend of his who was involved in nuclear research. I don't remember who this person was, but I remember standing in front of a big window that had multiple layers of high lead glass behind which were two mechanical arms that I could control from my side of the window. It was explained to me how deadly the atomic particles were and how we had to be extremely careful with atomic material as it was highly radioactive. With experiences such as these and as a student interested in physics, chemistry, and mathematics, all of this was very interesting and stimulating. My father had introduced me to the hard sciences and I have been interested ever since.

Unfortunately, my father became seriously ill from asbestos poisoning when I was eight. For the next two years, he deteriorated in health until his death, which happened when I was ten. At the time he was diagnosed, he was given six months to live. Luckily, he lived one and a half years more. During this time, my mother worked to support the family, so it was my responsibility to look after my dad when I got home from school. His skeletal face haunted my childhood for the next twelve years. It sent me into a deep childhood depression, which affected my ability to concentrate and work in school.

My widowed mother decided to move my sister and me to San Francisco in the hope that we could then afford a college education since she worked as a secretary, which didn't pay much in the 1950s. The move further isolated me from my extended family on the East coast and disrupted the family connections. My mother had told me how my father had wanted to see me become a physicist, but my life had been greatly interrupted. Growing up in San Francisco had its benefits. I enjoyed myself by going to the Golden Gate Park and visiting the museums. I had a paper route and took tickets to the Saturday matinee at a local theatre. I was a latchkey kid and made my way around the city. At one point, I was befriended by a pilot boat captain who would let me sneak on board and once the boat was past the warehouses, would allow me to steer it.

In San Francisco, we lived in a small one-bedroom cottage behind an apartment building. I had chores to do at home such as getting the laundry done at the laundromat, cooking dinner, and sometimes do a little shopping. Homework was not hard and I excelled at math. I had time to read and to think. My parents were well-read and had in their own way trained me to be a critical thinker—to question things and create assumptions.

There was a doubt in my mind as to the real wisdom of nuclear energy being a universal solution to the world's need for energy. The technical skills and education needed to operate a nuclear power plant are considerable. How could poor countries afford it? More importantly, how could we as a civilization justify the expansion of this dangerous technology around the world when the political institutions then and now were or are not stable enough to allow for the assured safety of the said technology? Was there another way for us in this world to go forward with the safety and assurance of clean energy for all? This was the question I set forth to answer.

Chapter 5: Something New in Energy

In May of 1972, on a lovely spring afternoon, while sitting on the deck of the log float in front of my houseboat that I had half-built from mostly reclaimed materials, I had what my aunt describes as an "epiphany." It was a geothermal-solar thermal hybrid power system integrated with other characteristics of cities and towns so that you would be able to feed it with municipal waste—methane from sewage that is processed. The ensuing thought process of designing this integrated hybrid system proved to be prescient for a variety of reasons as well as for one of my future projects, but at the time, I did not know this.

Let me explain the geothermal-solar thermal hybrid system. The concept is to take geothermal energy from the earth that has not become too hot and which can be used to create flashing steam and run a geothermal turbine and make power. Alternatively, we can take geothermal energy that is so hot (about 350 °F) that it can run an organic Rankine cycle generator (NORC). The idea is to take the low enthalpy and boost it with solar thermal energy created from a solar trough collector. Adding the solar thermal heat to the geothermal energy can create enough of an energy boost that it would go beyond 350 °F. In fact, it could go beyond 450 °F. It could go to 500 or 550 °F. It would run through heat exchangers and change the geothermal fluids into steam which would then produce power. The majority of geothermal energy is not hot enough to create power thus the need for more heat input.

Why would somebody think of this? It turns out I had been trying to figure out what other forms of energy we could use that would be long-lasting and safe for the environment—non-toxic, non-poisonous, non-radioactive, and does not emit anything into the atmosphere. As I was doing research for years, having looked at the solar science and having an understanding of geothermal, the idea came to me in an afternoon while sitting in the Multnomah Channel, a distributary of Willamette River in Oregon, and it really was quite a revelation—the old idea of "the light bulb came on" really seemed to have come true. I thought about this for many hours and went back to my research books. I started to create the model of the concept and write the mathematical formulas to calculate the thermodynamics of this system but I did have the research materials and appropriate scientific textbooks to give me the information I needed to create the formulas and to supply the energy balance information to work on. And so, I began to work on this concept.

After considering it and thinking about it for a couple of years, I went in during research to a patent attorney and asked what he thought of this idea. Interestingly enough, he thought it was original. He told me he would do a little research, and as it turns out, there was nothing like it in the patent office. It differs from the solar power tower system that creates a higher heat. The hybrid system takes advantage of the benefits of a linear-concentrating parabolic design that can be less expensive to boost the lower temperature geothermal that is not hot enough to drive a generator. The Patent Office issued me the basic patent for this process—Patent Number 4,099,381. From that, I began to think about how I was going to bring this concept to life. I worked on figuring out what it took to put it into play. I met with one of the senior engineers, the chief engineer of the Bonneville Power Administration, in about 1978 or 1979. He was also a professor of electrical engineering at Oregon State University. I sat with him a couple of times and discussed this concept and the idea. See the illustration on the next page.

I was aware that there was a place in Idaho called Raft River, a known geothermal resource area in southern Oregon called Klamath Falls, and a geothermal facility in Lakeview. Also, there was The Geysers in California. I began to identify sites in the Pacific Northwest since that is where I lived. I knew about The Geysers, but that seemed pretty far away and it was pretty hot. It was hot enough as it had flashing steam, and then the geysers, of course, became a major source of geothermal energy in northern California. I kept on learning as I went. I had written a proposal for research and worked with the Bonneville Power Administration (BPA) and the scientists at Raft River, Idaho. I made a proposal for which the Department of Energy awarded me an RD&D grant during the Carter administration. Around this time, I had also been introduced to a gentleman who used to be the head of the BPA, and who was familiar with the power industry in the Pacific Northwest. He aided me in getting a meeting at the offices of Pacific Corp in Portland from which I received what was one of the very earliest power purchase agreements (PPA). A power purchase agreement is a contract between a party that generates electricity and another party that wants to buy the electricity. The help with the utility came from Don Hodel who later became the Secretary of Energy. To see a video about the system, go to YouTube and enter https://www.youtube.com/watch?v=15zvBNBWUvI&t=68s. This should lead to a link for the video.

This power purchase agreement, typically called a PPA, comes from PURPA, the Public Utility Resource Policies Act of 1978. The act mandates that utilities must buy power from small power producers. At that time, the avoided cost for electricity that was being contracted for was based on nuclear energy. The Trojan Nuclear Plant was the most recent power plant that had been built, and so the rate structure for my little power plant concept was going to be based on the cost of nuclear energy at the time which turned out to be a feasible number.

As I was working on engineering, I was introduced to a large engineering firm that was interested in what I was doing and intended to make a little money on it at the same time. I started working with a senior engineer at that time, a wonderful gentleman named Rod McGathen, and sat with him to design this power concept into an actual blueprint of a working power plant. While we worked on the engineering, we waited for the funding agreements and the next funding disbursement round that had been approved. I also engineered and designed a woody waste power plant—a biomass waste power plant. At the time, it was about 20 megawatts (MW), but that is for another chapter. Prior to working with the engineering firm, I worked for several years on the Trojan Nuclear plant where I gained detailed knowledge of the design of the plant.

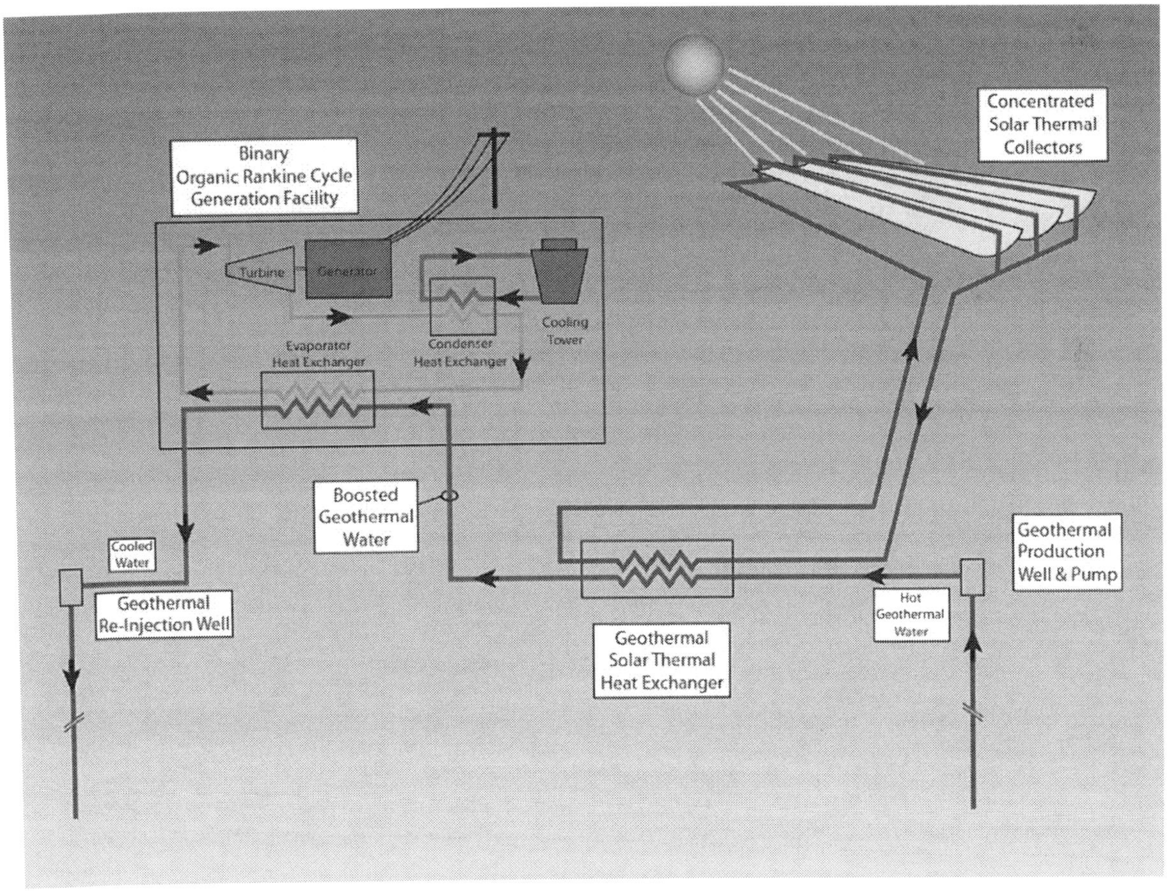

Figure 1: The Geothermal-Solar Thermal Hybrid System

As we waited for the approval of funding in 1980, an election occurred that ended with Ronald Reagan becoming the president. He canceled most of the research at the Department of Energy after being elected President of the United States. He took the solar panels off the White House as an example of his rejection of renewable energy and made clear his ignorance of the importance of addressing climate issues early on as well as for forms of energy independence.

I was to wait a long time for the hybrid concept to become a reality. In fact, it was only in 2015 that I saw my concept come to fruition with a Stillwater, Nevada utility implementing it. It was a fix to their 30 MW geothermal power plant which was underperforming at the time.

When I recommended the solution to that problem in 2012, they used my hybrid concept with solar thermal systems to address the underperformance of the geothermal facility.

In May of 2016, the front cover of *Power Engineering Magazine* featured aerial footage of just this power plant. In the magazine, there is a wonderful article about how this hybrid system is producing clean energy baseload; in this case, the plant is now a triple hybrid having photovoltaic arrays, a geothermal power plant, and a solar thermal trough to boost the geothermal to increase the production capacity of the geothermal plant and mitigate the degradation of the heat loss at that plant site. It has taken all this time to become a viable commercial-scale facility.

I believe these power plants should easily run fifty years with zero pollution where the spent geothermal fluids are reinjected into the ground at the aquifer level from which they came, which is pretty deep. It is well below any surface aquifer of any waters that people would use for potable water—thousands and thousands of feet below it.

One of the things that the United States Department of Energy has been doing is it has been looking for very high-temperature thermal inclines at depths of over 10,000 feet for what is called "enhanced geothermal systems." Most of what they have discovered except for one place, I think, is when they go that deep and find a really hot spot, there is no water that they can use, whereas the concept that I am advocating has more accessible sites at less depth and less cost in which the hybrid system mitigates this risk of development and front-end cost substantially.

The mitigated cost means that low-temperature resources in geothermal that are much more widespread, especially across the western United States, would present communities the opportunity to develop the hybrid system as an "eco-energy park," as I like to phrase it. This would mean that not only can the facilities provide power, but they can provide hot water through heat exchanges that can be used for greenhouse heating, aquaponics, hydroponics, and a host of process heat uses for commercial and industrial purposes.

As an illustration that this is entirely feasible, let me give a couple of examples. The most prominent, successful example is Reykjavik, Iceland with its hydro resources that are 100% entirely renewable. In fact, I have seen photos of a community outdoor pool which is all geothermal-heated and you see people in the middle of the winter swimming in the geothermally heated pool. They heat their houses and they keep the streets and sidewalks melted from the geothermal resource they have. Another area that I have already mentioned is The Geysers at Northern California which produces a good deal of power—hundreds of megawatts. Another location is Waiariki Field in New Zealand. Another area is Lardarello in Italy.

There are geothermal facilities now in well over 20 countries across the globe. They present multiple opportunities not just for power and hot water, but for applications in communities as a clean source of energy.

Chapter 6: Integrating an Organizational System for Sustainable Forestry and Biomass Energy

This chapter is mainly based on a paper I wrote and published years ago. It is self-explanatory and understandable for people not having any scientific background. An update for the current costs of fighting forest fires as compared to what the costs were when I wrote the paper is now according to reinsurer Munich re-estimates stating that the costs for all the wildfires that year (2018) is to be over $20 billion. So far, this year's fires should bring in a similar calculation.

Today, the application of an integrated model for sustainable forestry and biomass energy systems, which is an optimum organizational structure, will be the one I will continue to try and clarify for you.

Firstly, given that the forests have been experiencing drought, beetle kill, and catastrophic fires, there is a need for a national program that will address these issues to mitigate catastrophic fires and preserve the forests in a healthy condition for long-term sustainable yields. They must also manage the understory ladder fuels as well as promote the removal of the materials that cause catastrophic fires and the absorption materials for nurturing healthy trees over the years. As they are a healthier forest stand, this is better for water treatment. The lowest cost-effective manner of creating clean drinking water is a healthy forest or healthy watershed.

Secondly, a healthy watershed and healthy forest mean animal habitat is preserved and that animals are healthier in that environment. Birdlife will also be enhanced in that environment; there are multiple benefits to having healthy forests. However, to do this, we have to pull resources out of the forest that cause catastrophic fire, which in part has been a program of neglect. There is a huge part of the area that is impossible to treat without a major program being implemented in those areas and regions where the forests demand treatment.

One of the successful examples of where environmentalists, the logging community, and timber community actually sat down and discussed what could be the possible, workable solutions for all parties was called Quincy Library Group in the Sierra Nevada in northern California. This was some years ago, but it provided a successful model at the time, which, however, did not expand very far.

Forest service contract requirements have been evolving where the treatment is a form not of primary logging, but of a sustainable stewardship model. When we have that available over a longer period of time, the contract can then work to facilitate the development of a bioenergy resource plant. A minimum of ten years would be needed for an effective stewardship contract. Here again, there are some new technologies that are coming to the fore that I will explain in another chapter—how do we use biomass to create biofuels from waste streams of biomass? However, it will be tackled after further discussions on what kind of organizational structures would become successful models. Devising a successful model means you do not implement these plants everywhere; instead, you have them strategically located. There has to be some strategic distance between the placements of the plants, such that the resources are not depleted, but are in fact carefully managed for over many decades.

One of the examples that we have for carbon storage is wood since that is what trees do. Building houses and buildings out of wood has taken a step forward with a new type of material. A sawmill in southern Oregon is now making material from a German technology called "cross-laminated timber." The process is made from a very strong material to build modularly and is cost-competitive. There are now two buildings being built using this structure in the city of Portland.

Here again, this takes an older tree, not old-growth, but it means that forests will be managed for a long-term cycle for additional growth, which helps store more carbon and would make wood a stronger substituting material to concrete. In the milling process, concrete uses and produces more CO2 than wood. Additionally, there are new technologies on the horizon that are being introduced in the market to make construction feasible and price-competitive using the wood material for more carbon storage. These characteristics could be priced in order to offset the cost of treating the forest in the long term. Foresters will have to selectively think and manage vast acreages for the best use of a healthy forest.

The following paper is for a more detailed examination of the subject on integrating a long-term forestry management policy:

Integrating Organic and Organizational Systems for Sustainable Forestry and Biomass Energy Development

Marc D. Rappaport

Abstract

The development of cellulose to ethanol technology has been under research and development for a number of years. At the NREL labs, the Process Development Unit (PDU) has proven the technology. Commercialization of this technology to a larger scale is an important and vital step toward solving alternative transportation fuel production, global warming problems, air quality issues, urban waste disposal, and rural economic development. This paper presents a structure for scientists, business interests, and communities to use for building commercial-scale plants. The paper sets out the general functions of interested parties and how they meet their respective requirements for feedstock, financing, and risk management for a successful project.

Keywords

Sustainable forestry, biomass energy, ethanol.

Community Benefits

When developing a facility that can take such time and resources to bring it to fruition, it is beneficial to have community support. The new technology is one that very few people know which means that we must inform the public about the process and describe it in the simplest way while extolling the virtues in terms that bring the benefits home to the immediate community. It is not common knowledge outside of this group that fuel cells installed in buses are able to run on ethanol, producing very low emissions. In areas where the growth has moved to the forest edges, urban forest interfaces have tremendous catastrophic fire potential. This technology can utilize the residues from intensive management of the forest urban interface and should be roundly embraced in communities that face issues including waste disposal problem where too much of the landfill's material is already laying and the problem of burning due to pollution from pile burning is continuing. Fuels and fire management requirements for the reduction of residues are important firefighting issue that are without safe disposal alternatives. The seasonal job development from the increased programs of fire prevention with this process gives rise to job possibilities for those who are underemployed, students, or even prisoners. These programs would be separate from those activities in the wood products industry and would not be in competition for the same resources—an important piece of information for the wood products industry to understand. It is also a benefit to that industry for the new technology being available for a healthier forest products industry.

Many are going to say you can't prove it; however, global warming has a huge effect on weather patterns. We can't prove that the severe extremes in weather are due to this factor, but we can say that this technology will help reduce carbon dioxide emissions. The EPA has looked at the potential for biomass energy systems and concluded that they can be very beneficial for the environment. The importance here is lost to the congress and the public because global warming is not seen as was the atomic bomb—catastrophic in a direct, responsible way. Greater support from the public can have an effect on the political will in Washington and elsewhere if the public comes to understand the consequences of inaction around carbon dioxide emissions and global warming as well as the benefits of the biomass ethanol industries' potential. One area of direct consequence of ozone depletion is the increase in skin cancer—this is the kind of direct result atmospheric problems have on all of us when we fail to recognize the impacts.

Problems in Commercial Development

In the early 1940s, the Defense Department had Georgia Pacific build an ethanol plant at its facility in Bellingham using cellulose from wood as a feedstock. The plant is still in operation and produces 3 million gallons of ethanol a year. Simplot Corp. uses waste from a potato processing plant to produce ethanol. Cargill and ADM produce ethanol from corn. Today, over 1.3 billion gallons of ethanol are produced in this country. Why is it so difficult to build a scale-up of the enzymatic process for producing ethanol?

A financial analysis of such a plant has been looked at by NREL. Its analysis shows a return of about 20% possibility. In comparison last year, many mutual funds returned 30% with little work or risk on the investor's part. For the process that combines hydrolysis, enzymatic digestion for fermentation and distillation, a new approach to the risks involved for startup technologies is needed. A $75 million investment needs to meet many more requirements for satisfying investor risks. Lee Lynd, in his discussion of the near-term prospects (Lynd, 1996), points out that there are key costs to any such projects: "If the cost of any one of these is unusually low . . . the overall cost may be significantly lower."

This paper suggests opportunities in sustainable forest ecosystem management that have been thought to be too expensive for the Forest Service or timberland managers to undertake, but when seen in the light of the presence of the emerging biomass ethanol program, there are viable alternatives to be used. The cost component associated with the feedstock can account for 40% of the cost of production (NREL, 1994). When the cost is part of a program for biomass energy farms, the cost has to be borne by its end-use. When the feedstock is associated with solving a problem for the forest managers in long-term issues of forest health and wildfires, there is another dimension to the issues that bring about a greater symbiosis for the two different industries. Often, when we specialize, we may be unaware of problems in other disciplines.

Sustainable Forestry for Feedstock

Overstocked forests are not healthy forests. Diseases, drought, infestations of pests, and catastrophic wildfire are the results from unhealthy forests. Thinning stands for healthy forests take time and money. A commitment to the long-term multiple benefits from healthy forests becomes a different paradigm. Clear-cutting may be part of the tools used, but only on a smaller scale. Harvesting of timber comes from the focus on long-term benefits to habitat like cutting trees before or after the bird nesting season and stream enhancement for fish habitat and spawning; this change means a greatly enhanced need for intensive management of the forest. When this occurs, the results call for increased removal of brush and understory growth. Selective pre-commercial thinning becomes a greater requirement of forest management and with it comes increased cost and waste residue. However, there are costs for fighting wildfire and for prescribed fire. Catastrophic fire can cost from $1,200 to $1,600 per acre. Prescribed fire can cost $40 to $60 per acre (Oregonian, 3/19/1997). Residues on National Forests in the Northwest have volumes from 30 to 100 tons to the acre that needs to be removed. The problems of the forester in fuels and fire management are not what we would find if they were a molecular biologist. For many years, the Forest Service has fought forest fires. Today, in various National Forests, firefighting has contributed to the problems of the forests in the long run. Last year, fire burned more than 248,000 acres of Forest Service managed land at a cost of over $110 million in Washington and Oregon (Oregonian, 3/19/1997). There is a cost for disposal. Treatment is currently being carried out on selected tracts with costs paid for by the contractor if there is commercial material available for the Forest Service or if the site requires treatment with little merchantable material available. This is done on a bid-basis with short-term contracts.

Sustainable forestry has the benefits of selective logging with the long-term preservation of fishes and wildlife habitat, fuels and fire management being proactive, having effective insurance against catastrophic fire, and understory management being implemented as well as the benefit of becoming the feedstock for the biomass to ethanol plants that could be built in suitable relationship to National Forests and urban areas for early cost advantages. If properly approached, the process will yield a consistent output of merchantable timber and pulp chips as a result of the treatment of larger areas of land. Treatment would be an ongoing program that will not change with the fluctuations in the commodity markets. It is a program that takes into account the long-term view, and thus, it provides stable economic development for areas that have these plants. When these plants become established, it becomes feasible for the facilities to pay for the feedstock on a cost-sharing basis. Prices for feedstock could be tied to the rack price of regular gas; thus, providing the banking and business community the risk management of pricing against failure. Local communities would have a continuing source of waste disposal, jobs, and a renewable alternative fuel for oxygenates that are produced locally and sold elsewhere.

Overcoming the hurdles of feedstock commitment from the Forest Service on a long-term basis will be a significant step in assuring the long-term success of any such projects in the near term. As an agency that has its own mission and budget, it is not an organization that is in need of the ethanol industry. Rather, it is an organization that faces a changing environment and feels the need to adapt to it for it to benefit directly, an organization that wants to participate in such a commitment for the future, and an organization that is part of the biomass to ethanol industry where waste from the forests fuels our future.

Creating Capital Benefits

In starting this industry, the technology faces a number of problems dealing with the capital markets. In the past, there would have been greater support for the earliest development to receive federal funding. The recent battles in Washington are to balance the budget and reset the agenda of priorities. This technology has previously received some funding that have helped it get where it is today. Support for NREL has been strong, but it is hampered by the budget battles. This technology is one that holds great promise. In order for further commercialization, the addition of tax credits and accelerated depreciation for investment, specifically in the area of biomass ethanol plants from forest and urban residues, would be a strong investment incentive for the capital markets. This would be in addition to the retention of ethanol tax credits from cellulose.

The lack of familiarity by the banking community with the technology and its application for creation of the ethanol is another hurdle that needs to be addressed. Finding those areas of similarities with other industries that bankers know will help them understand the nature of the investment. The investment community would like to see a strong partner investing in this industry where they can also see a long-term interest in this process. It would be in the interest of the industry to show how the technology can evolve without outdating the hardware of plant facilities that make the ethanol. The enzymes and bacteria that feed on the cellulose are what can evolve through the use of the installed facilities. I think this is a critical piece of information that needs to be brought forward when approaching lenders. What are other uses for the plant's capacity to make a useful product? Are there other co-products that can be made with this technology? This technology is what Genentech pioneered. Why is it so hard to fund this particular application? One of the comments made by a venture capitalist at an NREL Forum was how can we be sure to be the low-cost producer if the oil industry wants to cut the market price? What happens if the tax credit goes away? The approach that solves the problem of feedstock supplies that are a problem for the Forest Service to dispose of gives reassurance that a facility will have the resource for a competitive product. The second question is one that can be answered with the addition of the tax incentives suggested above. The program worked for getting the biomass cogeneration industry off the ground in the early 80s.

Scientists and the R&D industry will have to work with technology suppliers and engineering firms to provide highly reliable facilities that can have a reasonable chance of coming in on budget and performing to specifications. Design construction firms that can give performance guarantees for the projects will be a key factor in which technologies become market-viable. Industry development will benefit from the support of a few major construction firms that make this technology one of their specialties. The ethanol industry is already well-established in its ability to make a marketable product although there have been some notable failures along the way. Careful attention is needed to avoid that part of the development in the early stages of this new process. The earliest plants must succeed to encourage the development of other plants. The niche market with significant cost advantages will be a prime starting point. BC International with its purchase of an ethanol facility from bankruptcy, in close proximity to a bagasse facility, gives this development some of the key important elements to succeed under difficult circumstances. Capital costs have been held down; feedstock comes from a close source and this keeps financing costs down. The cost of capital and debt will be another area for creative solutions to come to play when developing a profitable facility. This is where the role of local or state funding can play a big part in keeping costs down. Low-interest bonds, or tax-exempt bonds may have a role to play here. Local communities need to realize the long-term benefits from the development of this technology and give it the needed support.

Technical areas that are understood, but may have been overlooked in some facilities where contamination from the wrong kinds of bacteria can make significant reductions in ethanol output, must be adequately addressed (Kemmerling, 1989). The pretreatment of the feedstock can be a crucial step, having a major bearing on profitability if ethanol output drops by 20%. Drawing on the experience from the ethanol producers using corn, there should be a good deal of system operations that are well known to lenders. Drawing on that experience should facilitate financing. While drawing on the success of the corn ethanol industry, it would likewise be a stronger financeable project where the design engineering team has a track record in this field. Experience can save a project from future problems that add expense and could cause greater problems.

Organizational Chart

The organizational chart is a suggestion for seeing where and how the various groups that have a role to play in the development of the biomass ethanol industry come into play. Not everyone is included, parts are omitted due to space problems and in trying to show how the dynamics of federal policy can play out down to the smaller projects. The short descriptions following give a thumbnail look at what role that particular organization will have in the implementation of this industry and its projects.

A Conceptual Organization Explanation

1. The SED box represents the business development portion of the organizational structure. This organization will be from which the business venture will emerge. SED will undertake the commercialization of the enzymatic technology through the proposed structural diagram laid out on the previous page. The concept for this business organization is the work of Mr. Marc Rappaport, the principal of SED.

2. The MRI box represents the Midwest Research Institute. MRI has the management and research contract with the US Department of Energy. The role of MRI is to provide ongoing research for DOE and in the CRADA to provide SED with technological improvements to the technology for the production of ethanol from cellulose. MRI will provide technical support to SED

in the development towards commercialization. One of the goals of MRI is to commercialize the conversion technology.

3. USDOE is the Department of Energy that is charged with implementing national energy policy. In this structure, the goal of DOE is to see a large-scale commercial cellulose to ethanol industry develop to bring the dependence of the United States on foreign oil down with production of ethanol from energy crops. In addition, the goal of the US to reduce carbon dioxide by storing it in biomass energy farms is part of the strategy in addressing global warming and our requirement to meet the Treaty of Rio. To achieve these goals, it is the recommendation of this author that the suggested structure under discussion will allow the DOE goals to be met in the shortest time and with private capital. Since the federal budget and Congress have made no provision for building the first commercial facility of this type, it is left to the private sector to do this.

4. Box number 4 represents the first stage of organizational development to implement the financing of a commercial facility. This entity is the vehicle into which money, technology, and biomass development will be combined. In this stage, tax benefits will be apportioned 90% to investors, 5% to MRI, and 5% to SED Tax benefits comprised of ethanol gas tax credits as well as research development in the first commercial stage for R&D tax benefits to investors plus depreciation. These numbers should provide substantial early returns to investors. This stage will require the construction of the first commercial-scale cellulose to ethanol through the enzymatic process in the US about $75 mm.

5. Sustainable forest ecosystem management is a key element in the development of the earliest plants. In order to have a financially viable project, there will be a need for a certain level of feedstock availability at zero to a very low cost. Since the first-generation plants represent the greatest risk-wise, this process looks at solving an environmental waste problem as a solution to the feedstock problem. Working with the US Forest Service for the past several years, this author has the support and cooperation of the Forest Service for the development of feedstock for such a plant. A specific location has not been determined at this time.

6. The development and design of a plant that produces 30 million gallons of ethanol a year is part of this program. The sizing of the plant will be determined by a number of factors. Investor return, feedstock supplies, and the cost of electricity are some of the important variables. From the first plant, important lessons will be learned that can and will be transferred to other developments.

7. The availability of feedstock on a long-term basis has great potential incoming from National Forest lands. The Forest Service has a large problem with waste material on forests in the Western

US. Due to many factors, there is a large volume of material potentially available for such feedstock while helping the Forest Service solve an ecosystem forest health problem.

8. Once a plant has been built and operational experience has been gained, further development of this technology will be easier and more profitable. From known operating numbers, it will then be possible to build facilities with fewer problems and costs than the first. Through experience and engineering knowledge, the development company would move to license other plants on a royalty basis. Also, through our experience of marketing and engineering, with improvements from MRI, we will expand the technology where it makes economic sense.

9. The first-generation plants will be the riskiest to build on a cost-recovery basis. It will be profitable on paper without any operational history. In order to finance such a facility, it will be required that investors be financially rewarded for taking an early risk. This will require equity capital that is more expensive than normal investors would expect to see. Special attention in attracting this equity or venture capital will be required and negotiations will be important in the final outcome. Equity capital of 30% of the cost of the project would be a range.

10. Debt financial markets are required to raise the rest of the money to finance the construction and ownership of the facility. There could be two sets of institutions in this instance. A construction lender and the takeout or long-term lender would provide the 70% need for the rest of the financing.

11. Revenue for projects will be provided from the sales of ethanol to several buyers. Marketing of the ethanol will be an important element to assure the lenders that the sale of all of the output is feasible.

12. In the subsequent development, once a project has proven itself financially viable and the technology is working reliably, it will be a different set of circumstances under which future projects can be financed. Institutional investors who have a lower risk threshold and charge less for the invested money will be interested in investing in these types of projects. Here, we illustrate these institutions providing equity capital for future plants.

13. Since the development of the industry is in its infancy, there are probably technology suppliers that will be experiencing significant growth as we develop this technology. It may be very important to consider the strategic purchase of a key supplier to strengthen the company's future position.

14. Once the first-generation commercial facilities are built, there will be a period of re-engineering and design review during which it will become apparent what improvements can be

made to the technology and lower the cost of building these plants. We will want to have a major role in building several subsequent plants and training the personnel to run them. From this, the industry will have additional knowledge to make facilities efficient industry leaders and low-cost producers.

15-16. As expertise is established nationally and internationally, you can make the technology available in a package. Developers will be attracted due to the knowledge base and technology that is proprietary to these plants. At a cost that is less than their engineering and research would otherwise be, it should be possible to sell this technology and plant design packages that are at the cost of engineering and a royalty fee on a per-gallon production basis.

References:

Dombeck, Mike, (1997). A Burning Issue, Oregonian, Wednesday, March 19, 1997, In My Opinion. (Mr. Dombeck is the new Chief of the Forest Service.)

Kemmerling, M. Kent, (1989). Bacterial Contamination. The Fifth Annual Fuel Ethanol Workshop '89.

Lynd, Lee R. (1996). Overview and Evaluation of Fuel Ethanol from Cellulosic Biomass: Technology, Economics, the Environment, and Policy. The Annual Review of Energy and the Environment. Volume 21.pg. 403–465.

Chapter 7: Forests as Systems

One of the things we are going to have to learn is to see the forest as a whole ecosystem. The ecosystem might have watershed dimensions that can easily go beyond political boundaries; therein lies an additional complexity creating layers and layers of additional problems.

One of the things about the whole ecosystems and forests is animal habitat connectivity. Then, we have to deal with endangered species in that habitat. In the northwest, it is very clear that we have had to deal with spotted owls as an example as well as the marbled murrelet.

In the previous chapter, I tried to lay out an illustration that our ability to manage our forestlands on an ecosystem basis holds tremendous possibilities. Since the land, forests, and trees have their native habitat and have been growing in that environment for millennia, it is we who have the flexibility and the capacity to change our behavior and our institutional utilization of the resources for a higher, more valued purpose.

Let me discuss ecosystem management where the integration of multiple objectives will give more details. Ecosystem management could entail ecosystem management with concepts. Those concepts would have a management plan, management objectives, and then expected results in that ecosystem. Within the ecosystem, there is, what is described in the forest industry, landscape dynamics.

Those landscape dynamics include the layout of the forests, trees, and understory as well as checking the previous utilization characteristics whether it has had fire, has been logged and for how long, how long has it been standing, if it has been selectively logged for the understory, and those kinds of timberland management characteristics.

Within that, there is a review for whether there is diversity in that landscape—the variety of trees and shrubs—and these are also reviewed. The assessment of the ecosystem would also include the type of forests whether it is evergreen, deciduous, west coast wetter environment, or whether it is east of a mountain range like the Cascades or the Sierras where it would be substantially drier. Those are different broad-scale forest types and those have to be reviewed and examined to see what their natural states would be or would have been.

Within these ecosystem assessments, we have to look at the watershed characteristics or riparian functions. All of this then includes the kinds of birdlife that exist and the diversity of the bird species amongst the different forest canopies. Then, there are also active management strategies that we have to put into play so that we are assuming we have certain desired outcomes that fit the new definitions and the new goals for healthy forests, healthy habitat for wildlife, and healthy forests for water where the best filter system is a natural forest and healthy soils. All these come into play with riparian management strategies so that we have the desired outcomes for the use of the forest and for those who will be actively managing the forests for a long-term sustainable basis.

It is not just for the logging industry, but there are other resources available under the extraction on a carefully sustained basis. The sustained yield also has multiple goals to ensure carbon sequestration and tree health are optimized on a long-term basis. We may find if we let trees grow longer and not harvest them in shorter cycles. In fact, growing continuously they store more carbon. With such a practice, we have effective opportunities for carbon sequestration in our forestlands.

In recent years, we have seen some very significant catastrophic fires going across our forestlands, especially with droughts in California and elsewhere in the world. We have to make adjustments for this kind of catastrophic fire and find strategies that will help mitigate these catastrophic fires.

By mitigating the catastrophic fires, we improve the ecosystem's forest health. We preserve the viability of the soils and the ability of those soils to retain water and minimize the devastating floods that we have seen occur in more recent years.

This is going to lead to an important exercise to review the ecological, economical, and social values and find an integrated solution that works for the particular ecosystems in various locations across the globe.

Moving forward with an ecosystem management plan, we will have to take the social, economic, and ecological opportunities into play as well as some of the new technologies that are being developed in our national laboratories possessing the ability to add value to the many source materials coming out of these lands that need intensive management on a stewardship basis. A stewardship contract has to run at least for 10 years so that there is an opportunity for long-term planning and economic infrastructure investment that facilitates it with the ecosystem forest health benefits.

Using the knowledge and science from the forestry industry and silviculture, it is possible to understand the connectivity of habitat in the ecosystem. It is possible to have a much deeper understanding of the soil characteristics, of aquatic functions, stream runoff and fish health, nutrient cycling in this environment, and what happens when there are disturbances.

There is a tremendous amount of knowledge in our higher institutions of learning that can be brought to facilitate effective stewardship management. By combining these resources with new utilization technologies developed in our national laboratories, we will see the many health benefits forest protection will yield.

Chapter 8: Healthy Forest Systems

The development of healthy forest systems means that there will be multiple benefits for humans from preserving and protecting forests. Healthy forests can produce clean water, higher nutrients in the soils, sequester carbon, and preserve land. It can create a diverse environment for wildlife, stream health, and the mixed stands will produce varieties of species.

In this regard, what is going to be some of the opportunities that will derive benefits from healthy forests? With sustainable stewardship, forests can produce an ongoing supply of biomass. Reducing catastrophic fires also means that materials coming off of forestlands will not all be for timber. It reduces heavy fuel loading. This fuel loading, otherwise not removed, is one of the components that drive catastrophic fire. It also means that there are more nutrients in the water for the trees that will be remaining on the forest floor; the same trees will be able to use those nutrients to be healthier trees.

There is a variety of materials that these thinned materials can be used for. A variety of processes that I will disclose and discuss including discussions on how the national laboratories doing research in biofuels are creating a whole host of new opportunities for our society to have biofuels and chemicals—all produced from natural organic materials. There are distinctive capabilities in catalytic and thermal processing that can be applied to biomass conversion.

At Pacific Northwest National Laboratories in the tri-cities area of Washington State, they have biofuels in a biomass conversion research facility. This is a leading national laboratory managed by Battelle, received over 40 patents and numerous scientific awards, and contributed to a Nobel Peace Prize.

Biotechnology and bioprocessing are their main areas of focus. They are looking at an advanced analysis of materials that grow in the national forests and they are interested in looking into the sustainability of the natural forest for long-term benefits to the community. Again, this research in biomass utilization is led by Pacific Northwest National Laboratories.

In addition to Pacific Northwest National Laboratories' work, they also are doing developmental work in biofuels for aviation. They are looking at hydrothermal liquefaction, life cycle analysis, and fuel chemistry analysis as part of the methodologies moving the science of biofuels forward with a lower carbon footprint in the fuel and a lower fossil fuel content.

Another facility is Argon National Laboratories which is working on the life cycle approach, a life cycle analysis for biofuel production, and understanding the impacts of producing and using biofuels. They are looking at a holistic sustainable bioenergy landscape where productivity meets sustainability.

One of the comments they make is that their research shows that a holistic approach can result in production intensification. Nitrogen uses efficiency at the farm scale, restoration of contaminated water supplies, and mitigation of greenhouse gas emissions from both biomass and crops can occur with a holistic approach.

Idaho National Laboratories has a biomass feedstock system used at their facility. At this facility, they have a demonstration unit that can look at various fuels where they can use a hammer mill for grinding, rotary drying, pelletizing, cubing, and multiple packaging options. They study opportunities to be used from different feedstocks and how to reconfigure the materials to a useful state.

In their facility, they are transforming biomass into the feedstocks for combustion such as switchgrass which is made into pellets that are a quarter-inch, biochemical utilization of corn stover, thermal chemical or pyrolysis using ground pine, thermal chemical for gasification for eucalyptus, and waste energy creating refuse-derived fuels from ground waste streams. These are examples of what they are able to do at their facility for different feedstocks.

One of the pre-eminent national laboratories is Los Alamos National Laboratories known for its early research in nuclear energy and weapons. Today, it is moving forward with biofuels consortium and it is looking at reagents with grain fluorescent protein. It has a cooperative threat reduction program and is building genomics capabilities. It is looking into algae strains sequenced for biofuels and is doing advanced research in the genetics of plants.

They invented an acoustic flow cytometer resulting in an R&D 100 Award, and we have the advanced biofuels research facility in biosciences at the University of California Berkeley.

They have a new laboratory with new capabilities for the science of biofuels and biochemical having a tremendous capacity and are one of the leading research facilities in the country today. They can do enzymatic saccharification, fermentation, and a whole host of other processes to understand details at the molecular level, i.e. what is happening with materials.

The Biosciences Research Facility at Berkeley has a 10-year scientific strategic plan for advancing biosciences for the environment, energy, health, biomanufacturing, as well as additional capabilities in research. This, of course, is a tremendous environment for students who want to go on in graduate research for the science of biofuels and biomaterials for food and the environment.

It is a tremendous environment that is happening across the national labs for scientists to continue significant research so that we can have a greater understanding of how to care for our environment while it takes care of us. To that end, the forest service has also produced work that shows climate change can be mitigated through strategies in the forest and agricultural sectors by sequestering carbon in the national forests.

There are 170 million acres of national forest lands, many of which need to be treated. There are also many national forests and state forests that due to drought conditions and beetle kill, need to be treated, and as we have seen over the recent years, an extended fire season has burned up communities as well.

It is not just our forests in our communities. We only have to look at Alberta where they had a massive fire this last summer which got across and drove over 10,000 people out of a community. This is an area that needs to be addressed; it needs to be worked on. There are plans and abilities to address this that can actually create a considerable number of jobs.

With the new technology for biofuels, conversions, and biomass utilization, we have a manner in which we can decrease the overall cost. We will be producing higher-value products from waste streams, mitigating catastrophic fires and emissions into the air from those fires, and lowering small-particle contaminants being sent up into the atmosphere which only aggravates asthma for many people.

Healthy forests provide us with many benefits—mixed stands, diverse wildlife, better water retention, and filtration. The soil and the streams are healthier, and healthy forests are wonderful to experience.

One of the things we are going to have to address in maintaining our forests and having them healthy will revolve around ecosystem management. The ecosystem for forest health is something that has been going on in the forest industry.

Today's universities and colleges that have forestry courses have extensive information and it can all be drawn upon to create healthier forests given that we have current droughts occurring and beetle kill impacting large areas. It is going to be more important than ever that we have integrated programs for forest health in the long-term sustainability of our forests.

One of the major questions, however, is what mechanisms are there that we as a society can develop that will make this outcome possible. The cost of fighting fires has risen dramatically and the amount of acreage that has burned has increased. It is something we have to address since we have about 170 million acres of national forest in the western United States.

I have been thinking that one of the things we need to adapt to end the conflicts between the environmentalists and the logging community's desire for jobs and lumber products is to address sustainable forestry under stewardship contracts. It had taken more than a decade when the discussions began with the forest service to establish stewardship contracts.

Since the forest service typically gave out contracts that were for two years where the logging company would go in, harvest their logs, treat the area, clean up, and replant, of course, it was met with tremendous opposition, especially as we came to see on the old-growth forests in northern California—the redwoods. The redwoods were diminishing and you cannot replace a redwood forest any time soon.

There were changes in the law of the way the forest would be managed. To some degree, it has worked, but the drought conditions of the last several years have exacerbated the conditions of forests.

Oregon has relatively newer forest management laws where after every harvesting or when the amount of area harvested by clear-cut has been cut, it has to be replanted by those who harvested it. It has been working, but again, it does not address some of the larger national forest lands which are in need of treatment for fuel loading in the national forests in central or eastern areas.

This means that by the heavy fuel loading, this precipitates the catastrophic fires. When catastrophic fires burn through, they significantly damage the soil structure and the microecology of the soil. Hence, there are disadvantages for allowing fire—that too—not just natural fire but catastrophic fires that goes across national forest lands or other forestlands.

We need a program to really attack heavy fuel loading and address those kinds of catastrophic fires. It has been shown and demonstrated in tracts by the forest service that when these approaches to mitigating heavy fuel loading and fire come through, it is not as damaging as otherwise and the forest recovers faster.

There is documented experience and evidence that reducing heavy fuel loading can facilitate the management of healthy forests. The question really comes back to what is the cost of this and how many acres can be treated.

If we were to look down the road and say we are going to have a new structure so that we are really treating forest health as an important issue, we gain from significant forest health carbon sequestration and healthier streams, greater water retention, and wildlife.

All the benefits accrued are not necessarily monetary, but there is also a structure we can design as the managers of these resources. We can ascribe trees that will be left to grow on a longer growth cycle to store more carbon for carbon sequestration. Those trees and forests are designated for carbon sequestration and there is a value that could be an offset for dollars to pay. Some of the carbon sequestration, which after the appropriate period of growth when the trees are harvested, are then milled for a special cross-laminated timber construction method. It is then designated for carbon sequestration and has the offset of the funds that were allocated for the carbon sequestration to the additional costs of treatment in those forestlands.

Additionally, if we have the cross-laminated timbers which have been designated from carbon-sequestered timberlands, that carbon sequestration can have a value allocated in the housing or buildings that are built with this new construction method. Now the question comes down to the process by which we manage the forests; it is not the timber company undertaking a contract to log a given area, but it is a new structure that is given the responsibility with collaborative working teams to manage forestlands on a long-term, sustainable stewardship basis.

This creates an entity that is a facilitator to forest health and adjudicates the issues of timber harvesting from the timber industry. When timber and lands are managed on an ongoing basis, timber can be removed and can then be taken to a designated site where it is put in a log inventory as well as in the understory material which is removed to mitigate catastrophic fire and goes into a separate pile where some of it can be used for community firewood. The smaller wood material can be allocated for biofuels power and refinery technology that are emerging.

What we will be doing is designing and planning our social infrastructure to make it possible to have healthier forests that sequester carbon, clean water, protect the health of the soil and streams, and fish habitat at the same time. We have mitigated confrontations between the timber industry and the environmental community because it is incumbent on us to maintain economic health for rural communities on a long-term basis. However, on the other hand, it can be mitigated with the selective harvesting of the trees and not with the responsibility of the timber industry, but the timber industry and mills would have access to the inventoried trees from intensively managed lands for forest health and carbon sequestration. They also avoid the cost of logging.

Some of the ecosystem concepts for the management of this system include social, economic, and ecological processes that affect the treatment, and these need to be discussed and reviewed. Then, we can design certain management principles from which we derive the ecosystem goals providing actively managed ecosystems for the mitigation of infestation by insects and diseases by reducing catastrophic fire. We also can include sustainable practices as well as multiple species management addressing long-term health.

How does this happen? It happens in a management plan that identifies issues—social, ecological, economic—where communities meet and discuss these entire multifaceted and complex issues to develop the actual working plan that is compatible with silviculture science.

To address this, we would also be looking at watershed management objectives. You would have people who would be doing an inventory of watersheds, developing watershed and riparian management objectives, assessing the prescriptions for managing watersheds, identifying the mechanisms to implement those prescriptions, and then monitor the results over time.

This has been, to some extent, addressed in various forests and forestry colleges or universities. It can be implemented by simply redesigning the actual contracting systems and fire management regimens so as to address the millions upon millions of acres that need to be replanted.

The western forests, especially in the northwest, are different than the forests in the southeast which are different than the forests in the northeast and differ from the forests in the Sierra Nevada's. This is also an underlying issue about regional characteristics of the forests that have to be addressed, but it is something already known to the foresters who address these forest management issues.

In implementing such an ecosystem management plan, we would see additional recreational opportunities—more stable social and economic environments in the communities where resources are managed not just for one industry, but for the community as well as the forest health and ecosystem management.

Because of implementation as such, we would have the ability to sustain communities and there would be a more reliable supply of merchantable timber to the sawmill, which ultimately will increase the family-wage jobs in communities and will bring a more efficient use of public funds.

Simultaneously, it also diminishes the risk of natural catastrophes, enhances wildlife habitat, and improves the environmental quality, which altogether leads to healthy ecosystems. From this, there are also landscape dynamics that would be examined for each of the forest watershed areas. This leads to ecosystem assessment and active forest management strategies.

There are resources from science providing active ways to integrate successful forest practices for the benefit of communities while addressing job development, economic stability, and opportunities for new technologies to emerge in biofuels as well as materials.

One of the other parts of information to discuss is what is occurring in a number of the national laboratories around biofuels.

Having discussed some of the complexities of forest management for sustainable forest health, there is another part of the equation. People need to understand that the wood products industry is not going to be the major driver in what comes off those forestlands. It is what is in the long-term interest of forest health and the multiplicity of benefits that are active in the ecosystem management, and not just the manner in which we reorganize or treat the forestlands and the materials that come off of those lands and create a new interface.

On the other side of the equation, those materials that come off the forestlands are going to have to be processed for a higher value. Processing for higher value means value beyond the typical wood products industries—timber products and plywood or pulp and paper.

People have known for years that there are more characteristics and chemicals within wood than typically used. I am going to try and give you what may not be a complete list, but an overview of some of the tremendously valuable research that is occurring in the national laboratories.

One of those laboratories is Idaho National Laboratory and they have a focus on biomass feedstock where they have a national user facility. That user facility has opportunities for demonstrating to people who want to process material; they have hammer mill grinding, rotary drying, pelletizing, cubing, and multiple packaging options as part of their demonstration unit at the national lab.

They also discuss and get into supply chain development so they can assist people and companies with what is being processed and how to be efficient in the supply and logistics of the biomass fuel chain. Those include feasibility studies and economic assessments, storage performance characteristics, characterization of biomass resources, feedstock product characterization, and a fuel-supplied design.

Additionally, they provide assistance in processing feedstock based on the customer's specification. This allows sourcing for common and unique feedstocks, development of testing, and design processes. They can process to partner specifications on feedstocks and provide data sheets as well as packaging and shipping testing. They even go beyond that with feedstock specifications for biofuels, bio projects, and waste to energy projects.

However, this is just the one facility at the Idaho National Laboratories. They can also transform biomass into various feedstocks and provide illustrations for combustion from switchgrass, biochemical development from corn stover, thermochemical pyrolysis from pine, thermochemical gasification with eucalyptus, waste energy with refuse-derived fuels (RDF), and future commodities with palletization.

There are also Argon National Laboratories where they are looking at holistic, sustainable bioenergy landscapes where productivity meeting sustainability is a way to define their program.

At Argon National Laboratories, their approach while discussing sustainable bioenergy landscapes is that "through field testing and modeling, Argon proposes new productive landscape concepts to determine how it is possible to balance productivity and sustainability at the watershed scale."

They also discuss their systems that in evaluation can help watersheds reduce nitrate loading and sediment buildup while maintaining overall productivity. They have other goals, which are to lead watershed communities in designing a new agricultural landscape.

Along with those attributes in research, they also have a life cycle analysis for biofuel production and this software helps determine the lifecycle cost analysis. This is a program that is available for free; it simulates energy use and emissions of biofuel production and implementation over the full fuel supply chain. This is an important software for making an analysis of the development of bioenergy.

Another institution is Pacific Northwest National Labs. It is operated by Battelle Labs and is doing significant work in biofuel solutions for the aviation industry; they are working with Boeing and Washington State University. They have programs to do pyrolysis hydrothermal liquefaction lifecycle business case analysis.

Additionally, there is another research center doing cutting-edge technology where they are also able to show alcohol to jet fuel as a way to develop low-carbon, non-fossil fuels for aviation—an important development. They also have chemicals so they look at different chemical products that can come out of biomass feedstocks.

These are national labs with tremendous resources for advanced analysis, biotechnology, and bioprocesses with a focus on sustainability and have collaborations with other labs and universities.

From the PNNL, there is another national lab called Los Alamos National Labs and they are making contributions in biological sciences. In this environment, they are looking more at the atomic structure as well as the genomes of the biological feedstocks and how they are to be processed and used in the development of future applications, such as algae strains, reducing CO2.

It is a very advanced research program to add to the basic science and the basic knowledge of biological materials that will be the feedstock for bioenergy and biofuels.

I have some information on the vast biofuels research facility at Lawrence Berkeley National Labs. They have an advanced biofuels process demonstration unit that is for chemicals, analysis, biomaterials, and biofuels scale-up. Their biofuels equipment includes pretreatment like dilute acid, hydrothermal analysis, alkali, and ionic liquid. They have enzymatic saccharification, fermentation, and product recovery. They also do analytical chemistry and biology on a wide range of materials. This is one of the advanced research facilities at one of the country's premier research facilities.

Those are some of the examples of the exciting researches going on to determine the value and the materials that can be used from biomass and forest products beyond electricity or wood products for the timber industry and pulp and paper industry.

What is important to understand on this level is that these advanced technologies mean that the higher value from the chemistry in the forest waste streams will have a higher value to provide for being an offset in the costs of treating and maintaining sustainable forests.

Instead of having the forest industry carry the cost of logging and reforestation alone, sustainability is spread upon a wider base of higher valued products that this research is now bringing to the market. It would be these new technologies that will be placed in smaller rural communities having large forests, some of the national forests in eastern Oregon, eastern Washington, and northern California—those are millions of forests' acres.

When I was a member of the western governors' biomass task force some years ago, the task force assessed with the help of the national renewable energy labs and scientists, it was found that there was upward of 175 million acres that potentially needed to be treated and managed more effectively than had been done before.

At the time, some years ago, the western governors' association looked strictly at the value of the materials coming off the national forest lands, state forestlands, and BLM lands for power generation.

It became clear that they did not have a cost-benefit analysis that made it impossible to do that. There is no way that a singular biomass cogeneration power plant based on current avoided cost payments can carry any of the forest treatment costs. If, however, we are producing biofuels, then the value of biofuels, especially if it is like aviation fuel from biomass feedstocks, is considerably higher and that has the potential to make a contribution to those land treatment costs.

This is another value that has not been included in the discussion and the development has been delayed because the price of aviation fuel or natural gas has gone down as gasoline has gone down, and these startup scenarios are expensive unless they are undertaken by the Federal Government or with substantial government loans and supports.

It brings another concern into question—if we are going to treat the forests for long-term sustainable health, carbon sequestration, water cleaning, soil preservation, and carbon sequestration in the soils and root systems of the forests, how are we are going to have additional price support mechanisms that will facilitate these large-scale integrated solutions? It has to be understood that this is all for the benefit of largely rural communities and that it will create stable economic scenarios.

These will not be impacted by building cycles for wood products because you are going to have materials in the forests that are identified for long-term carbon sequestration; when harvested, they are identified as carbon sequestered feedstocks that go into what I described earlier as cross-laminated timber, and so, you will have more buildings built with timber that is stronger, more fire-resistant than metal, and has a lower CO2 output for its processing, so it is all in a long-term sustainable format.

Some of the new technologies beyond what is being done at the national labs, as one example, is of a friend of mine, Dr. Charlie Wyman, a scientist at the University of California at Riverside. He is working with a company called Vertimass and they are looking at high-yield conversion of ethanol into jet fuel, diesel fuel, and gasoline blendstocks. As Charlie explained to me, it is a single-step conversion of ethanol into a hydrocarbon with high yields.

It is a company using a technology that can be very transformative and price competitive making the bio-based fuels for jet fuel, diesel fuel, and gasoline blendstocks capable of moving into the existing supply chain while at the same time producing substantially lower non-fossil fuels for lower CO2.

It is hard to summarize the kinds and the number of opportunities and benefits that can be derived from biofuels and biomass. Looking at something a friend of mine did, a wonderful gentleman who has passed on, Bill Holmberg—Bill was one of the founders of ACORE, American Congress on Renewable Energy, and he was the chair of the biomass coordinating committee.

He had developed a list that runs to 52 different forms and areas of use of biomass. [Author's note: Maybe what I will do is reproduce this page in a book so people can see just how many opportunities there are for using biofuels. It is also that waste products are not necessarily dedicated crops, they are not exclusive to corn for ethanol—it is more diverse than that, and there are many forms of this biomass material.pres

There are a number of companies that are in the market of engineering these facilities today and that information is available through the Advanced Biofuels Association.

Another area that I mentioned as an opportunity for what develops in sustainable forests is carbon sequestration. One of the documents I have had for a little while is climate change mitigation strategies in the forest and agricultural section put out by the EPA in June 1995. In it is a discussion of forest ecosystems presenting particularly significant opportunities for carbon conservation and sequestration.

We have a review in science conducting this study background and it again looks at the carbon cycle and the forest carbon cycle. It explains it in some very effective terms and then starts to discuss the various scenarios for carbon storage, conservation reserve, wetland reserve, increased recycling, and reduction of national harvesting combined with a carbon sequestration program. The carbon sequestration program identifies forest trees for long-term growth which are then harvested and included in carbon sequestration-efficient building designs along with cross-laminated timber. Having that carbon sequestration, payments are offset for the landowners and those who grow the trees and are installing this new technology.

It is a good project to look at different kinds of models for the assessment of what is going to be an effective sequestration program; each region has different weather, water, and soil characteristics which makes tree-planting scenarios different from one area to another.

The regions that they reviewed were the south-central geographical area, the southeast, the north-central, the northeast, the Rocky Mountain, and the Pacific coast. Of course, on the Pacific coast, we have the Pacific Northwest and the Pacific southwest now.

They are projecting some fairly substantial carbon sequestration that could occur where they are talking about sequestration in the billions of metric tons of carbon and they had forecasted an ability with carbon storage and public and private lands to 37 billion tons and upward.

The dynamics explained in the study really go through both cost analysis as well as growth cycles, softwoods, and stumpage values, so it is not just a singular analysis, but it also takes into consideration cost constraints and cost demands. It should be considered something of value for us to draw on and I am sure that the silviculture forestry colleges have this material as well as the reference materials.

It is all science-based and truly valuable for what we need to do for carbon sequestration, clean water, and economic stability on rural farmlands in conjunction with the implementation of the new biofuels, bio refinery, and biotechnologies coming out of the national labs.

The Oakridge National Labs that produced a study a few years ago is entitled *1.2 Billion Ton Wood Waste Debris Available Annually across the United States.*

What is interesting about this study is that it reviews the available wood waste debris that is easily recoverable in towns, cities, and counties across the United States for recycling and not just for the recycling of power. To give an example of how significant the 1.2-billion-ton wood waste debris stream is in the development of the biofuels industry, a very conservative conversion would be that one ton of biomass can produce over 50 gallons of ethanol. I think today, they easily get 60 gallons of ethanol.

If you divide 60 gallons into 1.2 billion tons, you get a very significant number. I think it would be 20 million gallons of ethanol produced from wood waste streams across the United States. These 20 million gallons are something that can be supplemented to the RPS and the ethanol percentages mixed with gasoline.

Additionally, we discussed one of the technologies that is going to use a single-pass catalyst converting ethanol into biodiesel, and so, there will be some real advantages of this technology if we choose to recycle appropriately. Of course, biomass refineries need to have proper spacing across the country, so you do not want many clustered together, competing for the same resource. Each facility must have resources available from the waste stream at a possible low cost.

Chapter 9: Our Behavior—Whether It Is Conscious or Unconscious

We can choose our behavior, consciously or unconsciously, and we can set how we organize our society's values with education, both schooled and unschooled, on our social behavior with issues such as family and friends, jobs, money, and law.

Government policy and laws can adjust how we respond to changing our behavior. Today, cars come with seatbelts. It also took decades, but we now know that smoking causes lung cancer despite what the tobacco industry wanted us to know, so we have cut back on smoking tobacco which has cut back on lung cancer.

Today's most relevant problems and behaviors are texting while driving. It is, indeed, very dangerous and we now have laws taking actions for these behavioral patterns, continuously telling people to not text and drive at the same time.

In the same way, we now have a chance to educate ourselves and work on our behavioral patterns to lessen the harmful effects occurring on our national forests, forestlands, and the resources we derive from them—all because of the way we treat our environment.

While discussing this with my good friend, George, we realized that it is easier for us to change some of our behaviors to sustain our resources and energy practices at home like turning off a light switch and switching out light bulbs. It will be much easier than to quit smoking. It is easy to bring a change in our behaviors and less difficult than stopping ourselves from texting while driving.

We want to try and present an argument to encourage you, our readers, to realize that you have the choice, but you have to make it consciously and then continue to repeat it. You can do things like change the light bulbs to LED bulbs, replace the incandescent bulb, or even replace the fluorescent light bulb. It is not only easy, but it will also save you money that you can use otherwise on other necessities.

The next time you are in a hardware store, Home Depot, or Loew's, pick up some of these new LED light bulbs and install them in your houses or shops. An example is that one incandescent light bulb changed to an LED that can save about 500 pounds of coal in one year.

While pointing out the easy conversion of switching a light bulb from the traditional incandescent to LED, there are benefits both for the environment and for you. By simply changing the light bulbs, what you may discover is that it is available anywhere you live and there are sometimes coupons and sometimes trade-ins where you can get these new LED bulbs for free. You can just call your local utility and see what the energy conservation program is at your local utility.

I have been reviewing some information for quite a while and there are a variety of new technologies that are available as sustainable alternatives—some are not cheap to put in, but the light bulb is certainly the easiest and least expensive to make the first change with.

There are now eco-smart tankless water heaters, and the tankless water heaters save money because they only heat the water when you are using them, so there are only some related expenses for that as you switch over. There are hybrid water heaters and they combine a heat pump technology to preheat the water. This is just another way of you saving money and energy. Again, check with your local utility or Gas Company to see what technologies they are supporting and providing cost-sharing benefits on.

In Oregon, one of the companies that have become the standard is called Energy Trust of Oregon. It is worth your investigating little bit more about Energy Trust of Oregon as a model for states to use to create the offsets of CO2 reduction in conservation and weatherization, both of homes and commercial buildings, to offset CO2 emissions by a utility. You will see further information on how that program works because it has been very successful and it reduces the cost for homeowners pretty substantially in different technologies for conservation.

The other things that you can do is to simply go to your local Home Depot, Ace Hardware, or Loew's Corporation and look up information on where the gas water heaters or the electric water heaters are located, and there you will see a brochure that lists gas performance and the characterization of a platinum standard, a performance plus, or a straight performance; information will be at hand for you to compare some products that are currently manufactured for the home.

Along with those materials, you should find other information available on the US Department of Energy for energy efficiency in renewable energy on the website.

There are programs at your local utility and statewide programs in various states that will provide you with this information. Additionally, there are programs for renewable energy in states and there is information available online on how to find out what the status of conservation, renewables, and energy efficiency as legislated in your state is.

Some other examples of things you can find in the appliances for hot water include other technologies such as automated, state-of-the-art, and wireless thermostat controls that now use smartphone technology to monitor your house temperature as per your comfort settings. You get to set the temperatures and time, and it will run automatically.

They are very convenient, however, not cheap, but they save energy and money. There are several of those, again, that can also be found at your local hardware store, Home Depot, or Loew's.

Another simple approach to switching to newer technology is the use of weatherization tools—weather stripping for the doors, windows, and small cans of expanding foam insulation for sealing cracks. If you have an older house and it has cracks from the foundation or if it has cracks along baseboards or on the sides or doors where they do not meet the building anymore and you feel a cold air draft, you can use these cans of foam to fill these cracks.

In addition, there is a new technology company out of Portland, Oregon called Indow, and they make a hard-plastic window that is measured to fit your single-pane window. It pops in and out easily, but has a tight fit so it creates a thermal pane window effect without having to modify the windows at all—a really nice retrofit that is less expensive than putting thermal pane windows.

The other, of course, is adding insulating batting in the attic. If need be, you can add an attic fan so you can exhaust hot air in the summer. These are some of the easy weatherization solutions that are available—insulated curtains and cellular blinds.

In some cases, your older home may not have insulation in the walls and some companies will drill a hole in your exterior walls, pump insulation into your walls, and plug up the hole. In certain instances, you may want to check with your local utility and find out if they are available for assistance in the cost of weatherization from the utility.

Chapter 10: Targeted & Structured Use of Solar Technologies for Mitigating Deforestation

By: Marc D. Rappaport 5/2012

The ISSP being a cross-national collaboration program has conducted several annual surveys and events discussing a wide array of topics related to social sciences. The countries including the US, Great Britain, Germany, and Australia held each other's hands to lay the foundations of this organization and since then, the organization has expanded rapidly. Each nation participating has under their umbrella many academic organizations, survey agencies, and universities.

The recent discussion by the ISSP panel for the RIO +20 (a meeting in Rio De Janeiro, Brazil on 20th–22nd of June, 2012) meetings has presented us the opportunity to offer recommendations of the possible programs that can be used to solve or address problems of global climate change and poverty.

There is an opportunity to use GHG mitigation strategies with financial instruments like CDMs (Clean Development Mechanisms) when they are structured to a specific targeted program that can begin to create needed infrastructure for fabrication of sustainable technologies that can provide people with solar technologies for cooking & water purification, which will have a significant impact on village & tribal life for the good.

This technology is backed by many national and international organizations and has been tried and tested as well. There are several solar cooking tools, such as the parabolic reflector cooker, panel cooker, and box cooker, each of them having various designs with the same principle rule of cooking via solar radiation. Similarly, solar water purifiers use the means of UV radiation dismantling the DNA of microorganisms rendering them incapable of reproduction.

So now what I am proposing is that CDM's instruments should be used to raise the capital for the construction of fabricating plants that build concentrating solar cookers & solar ovens. These ovens & cookers would be built & packaged in a kit form, which includes tools for assembly & a couple of pots designed for the cookers. The capacity of the plants should be on the order of millions of units/yr. These plants could later add solar water distillers.

One of the benefits of solar energy is the modularity and variable sizing that can be engineered. I think that the universality of this technology can be used for expanding the development of peace across the globe. Creating work teams to solve disjunctive national relationship that are present in a competitive world can foster effective dialogue. To accomplish this end, I see value in organizing a team made up of different countries' engineers and designers to facilitate successful universal designs. To foster such development, I am putting forth a model to start with.

For the plant engineering which fabricates the cookers and ovens, I think it should be considered that a team is assembled that consists of Israeli, Swedish, and German scientists and engineers. The fabricating plants would be designed to be built & replicated in numerous countries. For starters, maybe Egypt, South Africa, Ghana, and Tanzania would be the countries in Africa that get the first plants. India, Indonesia, and Burma might get the next round of plants. China would have its own facilities to address the needs there. Haiti could be a location for a Caribbean plant. Expansion of these plants should be established on such principles in which they produce sufficient units for the plant's requirements.

The similarity in design of a solar dish cooker and a TV dish receiver illustrates that the growth of solar cookers could be an enormous number for use in high solar insolation areas of the world. A standardized design means interchangeable parts and a long service life for these products. The King of Saudi Arabia & the Sultan of Brunei could finance these units while receiving RECs for the energy offset. The CDMs would pay the costs of the plants. In various locations, reforestation programs would occur with the availability of the units. Locals would commit to not cutting wood for fires but instead, use the solar tools for cooking. They would be paid to plant trees and maintain them for longer growing cycles.

Chapter 11: Mind Games with Information—Media Overload and Apathy

What you know or think you know about renewable energy in the environment may not be entirely accurate. So, the decisions you make about the use of energy in many forms may be incorrect. The information in this book is an effort to provide you with correct information and resources. This will help you access the energy conservation and alternative energy uses in your community.

In charting the state renewable portfolio standards (RPS) for renewable energy, states set a standard so that utilities in the state are mandated to acquire a portion of their energy from renewable energy. This had come about due to legislation called PURPA or Public Utility Regulatory Policies Act of 1978. At the time, it seemed utilities were not very interested in acquiring renewable energy, and so, it became necessary for federal statute to require utilities to buy renewable energy into their energy systems.

At present, some states are enacting legislation that enhances the development of renewable sources while some states are not. What is your state doing about this matter? California, Oregon, Washington, and Massachusetts are some of the leaders in enacting renewable sources legislation. The city of Austin, Texas has also taken serious steps toward renewables as has San Diego, which has set a goal of achieving 100% renewable electricity citywide by 2035. The program in California is the most ambitious plan in the country and you can go to the California Energy Commission website for complete information. Hawaii has also voted for a 100% green energy plan.

A recent study showed that the top 10 states for solar electric capacity are Nevada, followed by Hawaii, California, Arizona, North Carolina, New Jersey, Vermont, New Mexico, Massachusetts, and Colorado. Some of this is due to legislation, while some of the states are facing internal battles for net energy metering which has set a manner in which people who have photovoltaic cells on the roof of their homes can collect a credit or in some cases, as payment for the energy produced on their roofs as the energy is monitored through a meter that can tell the utility how much that local facility has produced.

There are currently certain areas where Nevada has reversed itself and stopped net energy metering. This move has proven extremely contentious. The local public utility commission may have reversed itself as it saw the entire solar energy industry leave the state. There is a recent article that mentions that Arizona public service has reached a compromise in its net energy metering policy to work with people who have the systems installed.

There is a great deal of information available at the Energy Information Administration and can be found on the internet. Also, the US Department of Energy has supported energy efficiency in renewable energy division and that has a whole host of resources, information, and articles about the availability of energy efficiency for buildings and the programs that are available through the federal Department of Energy for conservation.

Renewable energy has tremendous economic benefits and this has been documented by studies in various institutions, including UC Berkeley. States and the feds have adopted investment tax credits to spur the investment in renewable energy, but to find out the exact status of a technology's investment tax credit of late, you have to check your state practices.

An area worth investigating is what your local utility has available for energy conservation and renewable energy. Many local utilities have weatherization programs for people with low income and elderly people. This includes financings that are fit within the monthly billing. Not all utilities do this, but it is worth investigating. Several utilities have exchange programs so you can take your old incandescent light bulb and turn it in and exchange it for a new light-emitting diode (LED). The difference is considerable. The utilities are actually giving out these LED bulbs when you take in an incandescent. Just check with your utility. The programs at your local utility may allow for your arranging to lease or finance a renewable energy solar system, a photovoltaic system for your house, in which case you need to talk to somebody at your local utility about the program that is specific to the location you live in.

If you do get to exchange your incandescent bulb for LED, you may want to pay some attention to the value of the light coming from the bulb. You have a choice, considering the intensity of the light. There are three-way bulbs, spotlight bulbs, and floodlight bulbs. Some are called "bright white light" and some are "soft white light." I have found that the bright daylight bulbs are very bright, leading me to replace and install six of these spotlights/floodlights in my kitchen. All six together use less energy than one of the regular bulbs that I had before. It gives a very strong, good light in the kitchen and lights it up very bright.

I have also changed some three-way bulbs in which case I looked for the softer light. Some people do not like the bright daylight white light. They are more comfortable with more modest lights, which is a softer light. It is something which you have a choice in and you may want to pay a little attention to that.

My comment about changing six of my kitchen spotlights out from incandescent to LEDs can also be illustrated as a 75-watt incandescent bulb replaced with an LED of equivalent lumens. Such a replacement can save 500 pounds of coal emissions in the atmosphere, which means if you change four incandescent light bulbs for four LEDs, you will be reducing the demand for energy and reducing the emissions by a ton of coal in a year from the power plant, which is very significant.

Cumulatively, if we were able to take an inventory of households that just changed four light bulbs, the positive effect on the environment could be extremely significant. This is one of the reasons you see changes in policy and the motivation to replace high-energy appliances, bulbs, and lights to the new efficiency bulbs and appliances.

There is no debate any longer; the fact is that if you pay any attention at all to glacial scientists, you will understand how quickly we are losing our glaciers and that the sea level temperatures are rising. The temperature at sea level and in the oceans is affected by the absorption of $CO2$, which is changing the chemistry of ocean water. $CO2$ absorption is increasing the acidification of sea water, which in turn is altering the health of ocean life. This raises eyebrows since oceans have been a stable environment for millions of years. We have seen starfish die-offs and various areas where the ocean has become a dead zone. These illustrate the critical stresses on the oceans that are leading to the loss of sea life. Another example is the emaciated bodies of whales washing ashore because they cannot get enough krill to eat as they normally would. The cold water welling upward that produces the nutrients for the small plankton and krill is insufficient as it has diminished over the years with the warming of the temperatures.

We now have more frequent experiences with *El Niño* and *La Niña* as the periodic warming in sea surface temperatures crosses the Pacific Ocean. They show global warming of the ocean very dramatically; you can see this in various NOAA weather satellite projections reading temperatures off the ocean. These are all illustrative of the true nature of climate change and its impact. How can you learn more about how to help the Earth besides this book?

There are several books that discuss the importance of reducing $CO2$ emissions and how you can save money at the same time. If you happen to have an interest in finding out more information about the reasons behind using less energy, you could check out a couple of books, the latest being Naomi Klein's *This Changes Everything*. Another one is *Plan B 2.0: Rescuing a Planet Under Stress and a Civilization in Trouble* by Lester Brown. There are also books by Bill McKibben, and of course, our noted vice president, Al Gore, *Earth in the Balance and An Inconvenient Truth: The Planetary Emergency of Global Warming and What We Can Do About It* (the movie had also received good reviews).

Other notable books include *Soft Energy Paths: Towards a Durable Peace* by Amory Lovins, *The Ecology of Commerce* by Paul Hawken, and *The Heat Is On: The Climate Crisis* by Ross Gelbspan—some very interesting books in the early reporting about what is occurring in the media and oil industry to lead people to understand that the question "Is climate change occurring?" was really a marketing strategy developed on Madison Avenue for the American Petroleum Institute not to entirely question climate change, but to raise a question to disrupt the progress of addressing climate change as noted by the IPCC report as though there is scientific debate regarding it.

My book list:

- *Earth in Balance* by Al Gore
- *Plan B: Rescuing a Planet under Stress and a Civilization in Trouble* by Lester Brown
- *Soft Energy Paths Toward a Durable Peace* by Amory Lovins
- *The Ecology of Commerce: A Declaration of Sustainability* by Paul Hawken
- *Economics for the Power Age* by Scott Nearing
- *Solar Heating Design* by W. Beckman & S. Klein, J. Duffie
- *The Heat Is On* by Ross Gelbspam
- *The Coming Age of Solar Energy* by D.S. Halacy
- *Energy Future* edited by R. Stobaugh & D. Yergin
- *Water and the Cycle of Life* by Joseph Cocannouer
- *Design for a Limited Planet* by N.Skura & J. Naar
- *The Step-by-Step Guide to Sustainability Planning* by Darcy Hitchcock & Marsha Willard
- *Sustainable Energy: Choosing Among Options* by J. Tester, E. Drake, M. Driscoll, M. Golay & W. Peters
- *The Sacred Balance: Rediscovering Our Place in Nature* by David Suzuki with Amanda McConnell
- *Sustainable Planet Solutions for the 21st Century* edited by Juliet Schor
- *This Changes Everything: Capitalism vs the Climate* by Naomi Klein
- *Cradle to Cradle* by Wm. McDonough & M. Braungart

- *The New Grand Strategy* by M. Mykleby, P. Doherty & J. Makower

Chapter 12: The Value of the Local Community

This chapter is very important because I try to lay out as we see from earlier chapters that the value of addressing renewable energy programs varies state by state. It is increasingly obvious and necessary for citizens to take a more active role in their communities so that alternative energy solutions are initiated in their communities.

We notice of late that California has adopted a 50% RPS, and there has been a bill introduced in the state legislature to make energy in California 100% renewable. Depending on where you live, there is much that can be done. However, this will require going to meetings of city councils and county commissions as well as having vigorous dialogs with lawmakers on a statewide level and at a local level.

You must become active in understanding what the person who seeks to serve the community represents and what they think. You will have to ask questions—very pointed and specific questions about that person's position on conservation, renewables, their perception of the oil industry's role as to whether that has subverted the dialog, and how astute they are at understanding science as well as how much value they place on science, proper engineering studies, and scientific studies in addressing these issues.

There are numerous articles written in many magazines and many publications on the value of energy efficiency and local renewal energy benefits. There are wind studies initiated state by state and studies on solar insulation which note the amount of sunlight falling on your state or even your locale. Some states, like Oregon, have numerous sites where they monitor solar insulation, so this is really good information available where you can look for what the amount of solar radiation value falling on your county is.

"The University of Oregon Solar Radiation Monitoring Lab (UO SRML) is a regional solar radiation data center. The goal of the lab is to provide high-quality solar resource data for planning, design, deployment, and operation of solar electric facilities in the Pacific Northwest. Creating the long-term solar radiation data set requires persistence, maintenance of high standards, and an effort to educate people on the importance of a solar radiation." [http://solardat.uoregon.edu/AboutSrml.html]

Geothermal Energy

Geothermal energy is identified in databases at the Department of Energy in the Geothermal Technology Office (GTO). Geothermal resources have been identified in California, so there are opportunities for geothermal development in that state. Some of the illustrations will indicate that Boise, Idaho has geothermally heated buildings. As the city's website explains, Boise has a river of geothermally heated water at its foothills. It is used to heat buildings, melt the snow on the sidewalks, and warm recreational pools. The city finds geothermal heating sustainable (See information on the GTO website). Klamath Falls, Oregon is also located in a known geothermal resource area (KGRA). Individual homes there are served with geothermal wells. Twenty-five years ago, the government there moved to heat several buildings in downtown Klamath Falls with geothermal district heating.

The Oregon Institute of Technology is run 100% with renewable energy due to its geothermal wells heating the buildings on campus. They even melt the snow on its sidewalks. The photovoltaic system they have installed also assists in offsetting the electricity demand. The Oregon Institute of Technology is 100% renewable, so such a thing is possible. Ground source heatpumps are highly efficient for heating and cooling.

There are other cities in countries like Reykjavik, Iceland that have huge geothermal resources. It is easy to look up and read up about Reykjavik and what Iceland has done with its vast geothermal resources. Geothermal resources exist in Italy and New Zealand as well. Numerous other countries have and are developing additional geothermal resources. In the United States, California has substantial geothermal resources—the Geysers is the world's biggest single geothermal field in California's Mayacamas Mountains. There are twenty-two geothermal power plants there with a total installed capacity of 1,517 MW. Nevada has geothermal resources that are being used.

One of the areas that I am going to discuss a little later is what I invented many years ago. It was an adaptation of low-temperature geothermal and medium-temperature geothermal boosted with a solar thermal system to bring the temperatures up, marrying together the geothermal and the solar to produce electricity, having it available for district heating, and re-injecting the geothermal into the ground so there is no pollution. There are applications for an electrical load and a thermal load in this system. I will discuss this a little later on.

Other Appropriate Renewable Technologies

Below are various renewable technologies that can be pursued according to what is appropriate for your area:

Windmills—the development of wind energy in this country and across the globe has been tremendous. There are instances in Europe, I believe in the Netherlands, of 5.7 million Dutch households that are powered by wind energy. The Netherlands on average generates 1.7 GWh of wind energy per day. In Denmark, wind power met 43.4% of its total energy consumption in 2017. Denmark aims to have wind power meet 84% of its electricity needs by 2035. An interview with my friend Don Bain who is one of the wind energy pioneers gives a personal insight into how wind farms come about. Wikipedia has an extensive article on wind power.

Solar PV systems—the photovoltaic systems in southern California and Nevada are very substantial and produce a great deal of power these days. Today, wind energy systems and photovoltaic systems can be competitive with natural gas, price-wise. According to the Energy Information Administration, the "levelized cost" of wind power is 8.2 cents per kWh. Advanced clean-coal plants cost around 11 cents per kWh. However, to put things in perspective, advanced natural gas plants cost around 6.3 cents per kWh.

Another form of renewable energy is passive solar building architecture. This involves designing buildings that consider the local environment while minimizing the climate's adverse effects on the comfort level of the building. This has been going on for decades. Placing south-facing windows in your building or your home can help solar gain and bring some light into the house. Also, simply adding a greenhouse to your south wall can help bring that heat in.

Small hydro systems—the definition of small hydro differs from country to country. Usually, you can take it to mean that they are below 10 MW in terms of installed capacity. These produce electricity with the gravitational force of falling or flowing water, and can be built in isolated areas where it would not be economical to tap into the national electricity grid (or where the national electricity grid does not exist).

In-pipe generation—there is in-pipe generation that is now being installed in water district systems bringing water at some distance into communities. Pumped storage is another form of renewable energy so that when there is excess power, water is pumped from lower level to a higher level where it is stored in a lake, and is used when it is needed. In-pipe water to wire power systems can operate across a wide range of head and flow conditions, and can be used in energy-intensive industries and irrigation districts to provide a consistent amount of clean energy. This is done without having to depend on the wind or the sun.

Cities use pressured piping grid systems to supply water where it is needed. Drainage and sewage systems are often gravity-fed. Both have untapped energy due to the abundant pressure.

Drinking water processors and industrial manufacturers generally install hydraulic devices to maintain a pre-set range of pressure to relieve the excess pressure and release it as waste heat. Using this waste heat is what in-pipe generation is about. All systems that use pressure-reducing devices could replace them with in-pipe generators while they maintain the same control on water flow and pressure, but at the same time, produce usable electricity. It is the flow through the pipes that is also used to generate electricity.

Building Design—another way to help the planet is by opting for a building design with energy conservation, regional architecture. In the Southwest, you have buildings designed using adobe bricks that provide for a thermal flywheel (a space or other element that gathers heat during a certain period, and releases it during another). This repetitive pattern of gathering and releasing heat helps buildings stay cooler in hot weather. None of this has to be too radical in design as there are designs that fit within the established architectural milieu.

Implementing policies that encourage district heating would be part of long-term planning for new power plants; we should see the development of eco-industrial districts or parks (EIDs or EIPs) with renewable energy as part of the system's design. EIDs or EIPs may differ from one another considerably; however, they collaborate in mainly four areas: energy, water, byproducts and services, and the exchange of waste materials. For example, instead of releasing surplus heat into the atmosphere, it can be re-used by the same plant or transferred to other district heating systems. Electricity generated from renewable sources can also be shared. In many instances, water is the heat carrier; however, water is a scarce commodity. Here, resource management can come into play with water treatment plants installed near water-intensive industries.

An illustration of how some of this can be brought to fruition is Oregon's 1999 electric-utility restructuring legislation. Legislation SB-1149 intended to establish a funding source for residential, industrial, and commercial electric efficiency, renewable energy, and market transformation programs. It required the state's largest investor-owned electric utilities (Pacific Power and Portland General Electric) to collect a 3% "public purpose charge" and give the go-ahead to the Oregon Public Utility Commission (OPUC) to direct a portion of the fund to an independent non-government entity. In 2000-2001, OPUC and other interested parties formed the nonprofit Energy Trust of Oregon. It has an independent board of directors and operates as per a grant agreement with the OPUC.

The nonprofit was charged with investing in cost-effective electric energy, implementing market transformation programs, delivering services with low administrative costs and program support costs, helping pay above-market costs of renewable energy resources, and maintaining high levels of customer satisfaction. The nonprofit is funded by five utilities whose customers pay a certain percentage of their utility bills to support renewable energy, energy efficiency, and various other such programs.

Another format is the state legislation in Washington. They created a solar initiative and a solar energy fund for the state.

The state of Washington is a proponent of sustainability and energy efficiency. Utilities there already produce most of the power used in the state via hydroelectric dams. The state offers residents and businesses many incentives, including net metering, tax breaks, and performance-based incentives. The state also has solar and other renewable energy resources, such as geothermal energy and wind energy. Washington's utilities offer extra rebates and low-interest loans for renewable energy projects. According to the Department of Energy's Energy Information Administration, "Washington has few fossil fuel resources but has tremendous renewable power potential."

Yet another way to conserve energy is through geothermal heat pumps. These should not be overlooked since they are very sustainable, clean, and can operate with low-temperature geothermal. People have laid pipes a few feet under the ground and use the ground temperature as a source of heat combined with a heat pump. It is very efficient and operates for the long-term; it feels very comfortable if used with radiant floor heating.

Another format in renewable fuels is the fuel cell. As we see the development of technologies that convert water into hydrogen and oxygen, hydrogen directly used in a fuel cell makes electricity and produces water vapor as a byproduct. This means zero emissions! Combined in weatherization in building design, we can significantly decrease the demand we make on our electric systems, lower our emissions, and still retain the quality of life that we are used to. "A lot of people walk the streets of Downtown Boise and have no idea that such a big footprint of Downtown Boise is heated with this renewable, clean resource that we take advantage of in a big way every day," said Colin Hickman, communications manager in the Public Works Department for the City of Boise.

In talking with a friend who has installed fuel cell power system, I asked if he would like to share some of his thoughts on them with me for the book. The following are his comments:

Fuel Cell Technology

"This is written for laymen who are interested and who are concerned with truly sustainable distributive electrical power supply. This is a quick review written for the pragmatic and operational realities of fuel cells. I want to encourage all who read this to drill down into the truth and bypass all the distracting and mostly false public relation rhetoric floating around the media. This subject is simple and straightforward and easy to understand if you choose to seek the truth. Please refrain from thinking only the experts can grasp and utilize this technology, absolute poppycock! Remember Farnsworth was a potato farmer from Idaho and he had a glass blower relative who made the first Cathode Ray Tube and then they created the Television. They were pragmatic novices who were not "experts" until after they created a world-changing invention. What ever happened to the American ingenuity needed to solve the unsustainable uses of fuel sources currently used in electrical power creation and the centralized distribution grids used today.

Experts are good for solving individual issues within a system, then you can tell them to go away after you pay them. After reading, you will hopefully find the key to unlock a truly sustainable and renewable power source. Here we go!

- Types of fuel Fuel Cells:
- Proton Exchange Membrane Fuel Cell/Polymer Electrolyte Membrane (PEMFC/PEM)
- Solid Oxide Fuel Cells (SOFC)
- Molton Carbide Fuel Cells (MCFC)
- Phosphoric Acid Fuel Cells (PAFC)
- Direct Methanol Fuel Cells (DMFC)
- Alkaline Fuel Cells (AFC)
- Reversible Fuel Cells (RFC)

Since we are advocating pragmatic solutions to utility power and transportation energy supplies, we will review only the first three. The others have only specialty applications and have too many faults.

Let us start with the basic operating system of the fuel cells. Simply, it is an electrochemical device that uses hydrogen and oxygen in a chemical reaction to supply electrical current.

Fuel cells all have anodes, cathodes, and electrolytes throw in diffusers and catalysts for good measure. The anodes, cathodes, and electrolytes are different from the different types of fuel cells. The devil is in the details on how you get the hydrogen and oxygen especially as we are focusing on a sustainable model and on trying to stay away from noble metal catalysts.

First fuel cell to be reviewed will be the PEMFC. This is first mentioned as Electrical Vehicles are all the rage currently. PEMFCs at 80 °C are low-temperature fuel cells and therefore have short start up and cool down times as thermal expansion is less of a problem. They are small and light and can be fitted to a vehicle. These EVs use pure hydrogen from a high-pressure storage tank and oxygen from the air. There are current commercially available vehicles on the road as we speak.

Due to the hydrogen fuel requirement, cost of producing hydrogen, and available hydrogen distribution fuel stations, or lack of these, PEMFC vehicles are not practical at this time in comparison to Battery EVs. Additionally, these PEMFC vehicles cannot take advantage of regenerative braking which lowers their total efficiency numbers in comparison to battery EVs. These PEMFC will most likely be used as onboard battery chargers to extend run times of battery EVs. Other negatives are the longevity of service, issues with flooding diffusers, and membrane humidity which still are durability challenges. Additionally, expensive noble metal catalysts are required. The positive is that the exhaust is mostly water as no carbons are used in fuels; at least, at the local level. Carbons might be used in manufacturing of the Hydrogen, so make sure only solar, wind, or hydro power are chosen in the hydrogen supply stream.

The second group of fuel cells, stationary units, are discussed next. These fuel cells are large in footprint and very heavy so they cannot be used in transportation. These units can be as large as 10 MW and growing larger every year. Molten carbonate at 650 °C and solid oxide at 1000 °C are high-temperature processes. The two advantages with high-temperature fuel cells are that they can internally reform natural gas into hydrogen and they do not need to use noble metal catalysts. The negatives of high-temperature fuel cells are the very slow start up and cool down times; therefore, they are best for always ON with continuous base loads at least equal to maximum rated output of the fuel cell. The high temperature also causes extreme thermal loads and corrosion within fuel stacks, especially the bi-polar plates.

40000-hour (5 years) lifespans are typical currently in the industry and are maintained properly and not heat-cycled excessively.

MCFC are 65% efficient with a turbine on the backside and 80% efficient when using waste heat in CHP process. SOFC are 60% efficient stand alone and can be 80% efficient using waste heat in CHP process.

Both these fuel cells are with high enough heat not to require noble metal catalysts which is a positive. The fuel of choice is currently natural gas due to the wide availability and distribution piping. Digester gas is used in the livestock industry as well. As all fuel cells require pure hydrogen, the reforming process uses high-temperature steam shown in flowchart below:

Steam-Methane reforming

$CH_4 + H_2O$ (+ heat) > $CO + 3H_2$

Water Gas Shift (if used)

$CO + H_2O > CO_2 + H_2$ (+ a small amount of heat)

Reviewing these equations, the sustainability process see that carbon is introduced into this process. Carbon is undesirable in any sustainability program. For this process to be truly sustainable, the carbon must be sequestered from the exhaust stream or used in another process that requires the carbon. Additionally, natural gas contains sulfurs which must be trapped in filters. Digester gas has much higher sulfurs and nitrogens which must be trapped.

As you can see, fuel cells using natural gas do have waste products that must be considered carefully. Highly filtered water is also a necessary product required.

Combined power and heat process (CHP) are desired to reach the 80% efficiency available.

Due to the statements directly above, Caveat Emptor must be applied when choosing a manufacturer. The big guy in the industry in the SOFC market falls short in warning issues.

I detailed above true CHP ability and waste products that are carefully not mentioned. Better to look at the smaller players that embrace the CHP process.

Molten carbonate fuel cells are large installations manifesting as a single unit with stacks within, constituting the fuel cell. A large footprint in a single concentrated area is required.

Solid oxide units are modular in nature from 25KW to 250KW per module and typically, therefore, can be creatively laid out to match available space configurations.

In conclusion, the fuel cell can be a valuable tool for the sustainability minded energy consumer. They allow microgrids that are the future of sustainable power. They can operate on multiple fuel sources with the proper treatment up stream. Fuel cells do not follow a diurnal schedule; they don't care if it is cloudy or if the wind is blowing and therefore, do not require power storage if base load to demand ratio is carefully calculated. When sustainable power sources are sufficiently available to electrolyze hydrogen and distribution networks are available, the dream of power production with water as the only input into the system and the only output from the system is realized. In the meantime, choose existing systems carefully and do not fall for the rhetoric used by the marketing efforts of the manufacturers. Dig deeper and complete your due diligence!"

Chapter 13: Sizing for Local Resources and Regional Needs

While considering the various technologies available, it is also important to consider the resources of the community, the size of the population, and the economic base that forms the community and the surrounding area. Is it an urban area with a large population or a small farming community? Would a modest community solar PV system be more cost-effective than individual systems on the houses? For example, micro-energy systems are something that villages in more rural areas that have little or no electrical infrastructure may want to consider.

If, for example, there is a large forest area and that forest is under stress (for example, it has succumbed to beetle kill, drought, or overcrowding), the risk of catastrophic fire can be averted by removing ladder fuels and cleaning the understory to gain optimum tree spacing. A solution can be found so that the woody waste biomass material goes to a bio-refinery which makes use of cogeneration. This can be combined with long-term sustainable forest management that focuses on forest health, carbon sequestration, and long-term local jobs for a sustained yield to timber mills that produce a new timber material—cross-laminated timber.

Structures using this cross-laminated timber that come from sustained yield carbon-sequestered forests could have some certifications proving that the timber is carbon-sequestered that had a long-term growth and that has some price advantages and tax credits that offset the potential longer term growth cycle that has delayed the revenue return to the forest landowner.

The above is a more efficient system that is just now emerging. I think there is development ongoing in Germany. In Portland, Oregon, there is a four-story building that has been completed, and they are working on a building that is going to be twelve stories high. Cross-laminated timber is structurally every bit as good as steel. If a fire was to occur, it would be more resilient since it would not have a catastrophic failure once the temperature reached the melting point of the steel beams. It would be able to resist more successfully in the sizing of the beams used for the cross-laminated timber as well as the structural value of the thickness of the floors.

As an illustration of combining carbon sequestration with advanced building materials with advance land management and forest management tools and policies, we need to segregate the mill industry from the forest management practices. The long-term value of preserving water quality, forest habitat for animals, and long-term growth of healthy trees that can sequester more carbon is ongoing with sustainable harvesting so the local economy is stabilized and not whip-sawed through economic boom cycles and recessions.

Another coming option is to use a bio-refinery where the application of new technologies that use cellulose as a feedstock for power, fuel, and chemicals and where higher value materials are also capable of being produced can interplay between the successful and advanced economy for sustainable forestry and sustainable bio-products.

In talking about the value of the local resources, another illustration comes from the Southwest and its high solar insolation. There, we have seen a number of very successful installations in PV as well as some high-temperature solar thermal systems. This is not just in the Southwest—this trend is occurring in several locations throughout the southern states.

Additionally, sustainable energy is not just solar insolation; there are opportunities in a variety of areas. The windy parts of the country are potential areas for additional windmill installation; thus, making necessary the retitling of the REA (Rural Electric Authority) to RREA (Renewable and Rural Electric Authority). Using the family farm as the base for windmills minimizes the pressure on large corporate farms to adopt windmill technology. Thus, the smaller family farm can have a commercial utility windmill on its site and use lease payments on the land to supplement the income to the farmer and help sustain long-term organic farming at the family farm level. This would be enormously beneficial for our country and for preserving the soils and water.

Another area to look into is large feedlots and dairy farming or pig farming. We have to use the accumulation and processing of animal waste through digesters and anaerobic digesters to remove the waste streams from surface water degradation and burn the methane in production of electricity. This arrangement can be used to offset some costs of operating farming facilities. The treatment of effluent in such farming operations is going to be enormously important.

Chapter 14: New Technologies and the Impact for Local Control

The use of solar PV is not new. It goes way back to the beginning of the space program—they were used in the space program to provide power to the satellites, but it is in the last few decades that its use has become more widespread. Today, the International Space Station is powered by P.V. Companies like Solar City and First Solar are examples of this growth.

Other technologies emerging in other fields like biofuels that have growth potential for communities with resources and that need to address biomass or forestry waste to mitigate catastrophic fire, beetle kill, and drought (that have devastated forest). These communities will need to be treated to reduce the size of fires and reduce smoke pollution in the atmosphere.

There is an emerging process called biomass gasification which converts solid biomass fuel into a combustible gas. Another is torrefaction which is the thermal process required to convert biomass into a coal-like substance that has improved fuel characteristics than the original substance. Yet another new technology is simultaneous saccharification and fermentation or SSF for short. It has been pioneered by the National Renewable Energy Laboratory in Golden, Colorado to convert biomass into conversion to sugars and then into ethanol. A new catalytic process in R&D takes the ethanol and convert it into biodiesel. Some technologies are emerging past the research and the bench scale and entering commercial operation.

As this moves forward, there is opportunity abounding across the United States and across the planet for the utilization of biomass waste. This means not removing whole trees but implementing utilization of waste streams. In the United States, the Oak Ridge National Laboratory did a study in 2011 entitled *2011 U.S. Billion-Ton Update: Biomass Supply for a Bioenergy and Bioproducts Industry*. This study concluded that there are over a billion tons of existing recoverable wood waste and agricultural waste that are produced in this country every year that need to be addressed.

If this existing waste stream that is already in the system were processed through new biomass technology-driven refineries, then there would be a tremendous new source of jobs and economic revival with very low or zero fossil carbon emitted to the atmosphere. This holds tremendous potential for biofuels, biodiesel, and biochemicals as substitutes for oil-driven resources.

One of the very prominent examples of new technology growing in the field of renewables is the advent of electric-powered cars and the hybrid cars. The Volt and Tesla cars as well as several others are coming to the fore. General Motors has now come out with the new 100% Bolt that gets over 200 miles in a charge. That is a remarkable feat! The automobile industry is prepared and has the technology to adapt to a low-emission and zero-emission transportation system—all to the benefit of communities and the environment.

Another new technology is fuel cells. Fuel cells run on hydrogen and can power cars. The developments have been on-going for a while but are slow to materialize in commercial space.

Some organizations are doing their best to bring new technologies to the public and make them more accessible. One such group is called Verge out of Oakland, California. The Verge group have put together a micro-grid to run 100% renewable energy to power a large part of their conferences for the last six years. There were a multiplicity of sources and the micro-grid development that they are working with includes companies such as Hitachi, Direct Energy, JOM Energy, and PG&E. In this endeavor, you had combinations of small gasifier power generators, windmills, photovoltaic, alternative fuels, and energy efficiency, all combined to provide sources of energy that help run the conference.

You can look them up and find out additional information. See if you can attend one of their conferences. You'll see independent technologies that are in island mode use or onsite generation that can lower your energy cost, improve energy reliability and resilience, and lower the carbon footprint of what is happening in your community.

At a Verge conference that I attended in September of 2016, there were a variety of companies that were selected—I am not going to go through all of them. Illustrative of what was to be found, there was one company out of Hawaii called Pono Home whose mission was to help people save money and make their world better. Specifically, its mission was to help people save money in utility bills and live a greener, healthier life. Their services were not just one-off. They could do an energy audit that included electrical appliances and suggest ways to cut down on the energy usage. They could teach you to identify ingredients in food and personal care that you may want to avoid. They could find ways to help you reduce your gasoline bill, save on fuel, and show you how to increase the tire pressure and even pump them up if you needed it. They could also help you remove gunk and dust from appliances and improve indoor air quality. These folks were just using conservation, weatherization, and a host of other ways to help people immediately benefit for their homes in a way that would save them money and make a better environment.

Another really interesting company was called Smart Yields (smartyields.com). It uses information in a software program that helps farmers. They could do field installation with sensors that monitor plants for the wetness of the ground, soil sensors, soil moisture, micro-climate sensors, and temperature and humidity. The sensors could also detect atmospheric pressure. They had different kinds of water sensors for dissolved oxygen, conductivity, different flows, leakage, and water temperature. They also had case scenarios for CO, CO2, CH4, hydrogen chloride and nitrous oxides, and liquid petroleum gases as they may affect the agriculture crop.

They could do other monitoring activities which would all then remotely feed to a computer at the farmer's location of choice. With the wireless information feeding into the farmer's computer, he would know precisely what is going on in his operations. This was a good example of using new technologies to help farmers become efficient in their operations and save costs.

Another technology is gasification power generation. The company that was at the Verge conference, ALL Power Labs from Berkeley, California, had affordable portable carbon-negative energy generation from biomass-powered gensets. That sounds like a mouthful, but it has been some years in development where using waste biomass, potentially nut hull remnants, and other similar materials can be fed through a hopper into a gasifier that produces what is called a "syngas" (synthetic gas). It is a lower BTU gas. That gas is then fed to an engine that uses the lower BTU fuel to turn a genset and create electricity. So, these folks have been working on this concept and have a model currently available. Their website is allpowerlabs.com, and they have quite a bit of material and information on the different sizes of the plants they have engineered, the various biomass feedstocks, information fliers on operations, and performance for communities that might need power that is off-the-grid or village-size.

Earlier, I mentioned that there are different policies, programs, and laws in place across the fifty states, and part of the problem is figuring out where your particular legislature falls in the spectrum of support for renewable energy or alternative energy systems.

Another group that I came across at the Verge conference was Advanced Energy Economy. They have put together what they call a "power suite." In this, there is a very extensive bit of software with a robust set of tools that includes a billboard, a docket dash, and power portal. The power suite allows you to track and collaborate on energy legislation and utility regulatory proceedings across the country with one easy-to-use interface. They have quite an interesting and extensive list of search characteristics and tracking and monitoring developments both daily and weekly that you can follow (to collaborate and synchronize your team's policies, summaries, positions, and priorities).

This is all designed to help you and your communities understand what is going on across the country and to help you effectively implement new policies or plans or collaborate across the country to build the response within your community for renewable and alternative energy. It has a membership fee—it is not cheap—but when I looked at what is available (there are memberships across the board) there are two kinds of memberships: commercial and nonprofit. You can subscribe monthly or yearly with different functions available at different prices. Look up powersuite.aee.net to find out what is available, and how they can help you.

Illustrative of some of the functions that you will find to stay informed are their blog, white papers, webinars, a public power portal, and several policy initiatives. They help you exert influence and they will help you with programs regarding direct legislative engagements, state partner collaboration, working groups, and business councils. They will also have information on federal and state policy activities as well as on clean power plan campaigns. They also have an area to help you build relationships to gain cutting-edge knowledge, information on new technologies, and how to influence policy changes. I urge you, if you are a member of an organization, whether it is public or profit-based, to look into these folks in different states. Folks in California are largely ahead of the rest of the country, but do look into those areas to see what help you can get.

One more example of information that I got from the Verge conference is GRID Alternatives. It is a nonprofit organization that helps provide low-income alternatives to renewable energy for low-income families and international access to low-cost energy. It also supports a STEM program (science, technology, engineering, and mathematics) to help people develop the skills to participate in the renewable energy economy. They provide assistance for tribal access to solar programs as well as for collegiate programs. I urge you to look into gridalternatives.org to find out what programs they have going to get involved in helping your community develop programs and actual projects that can support your community.

One such program is offered by SunPower, and another by Sunrun which provides cash and equipment donations to support their national programs, and lastly, M-Phase Energy helps grid equipment get installed. To date, they have installed 3.5 megawatts of solar for low-cost families. What is being done in communities today for others is done in part on a voluntary basis. It is also specifically done to help low-income communities afford renewable energy.

One more group I want to acknowledge is the national organization, Clean Tech Open Challenge Competition, which features regional competitions for new technologies designed to help the climate. They have now joined the Solar Impulse Foundation to help startup companies succeed.

Chapter 15: Turning Down the Thermostat

One of the toughest jobs we must all tackle is reducing our carbon footprint. Each one of us has a role to play. However small it may seem, taken together, it has an impact. One of the simplest, easiest, and tremendously beneficial things to do is plant a tree. Trees are one of the best carbon storage systems we have available. They are natural, and the opportunity to grow trees exists almost everywhere. You can visit the Arbor Day Foundation and purchase a tree for planting.

Forests will be cooler from the shading than other areas. If you live in an urban area that has lots of parking lots, streets, sidewalks, and buildings reflecting heat, you can tell the difference when you come from it to a forested area. It can be as much as 10 degrees Fahrenheit cooler in the forest from the urban environment. Trees are very critical for soil stabilization. They also help store carbon in the soil. You may have noticed that once a year, on Arbor Day, people go out and plant trees. Perhaps you can participate in your community and have a volunteer force that goes out and makes it a special point of planting more trees on the current Arbor Day than have been planted previously.

In planting trees, there is also a discussion of what happens and is happening in our forestlands. Living in the Northwest, I have become sensitive to Northwest forest issues such as practices for sustainable harvesting in the forest. This has been accepted as a process by the forestry department and the wood products industry seeking long-term sustainable forestry methods.

Addressing long-term sustainability successfully means we also have to be conscious of the climate change's impacts to the forest. For example, climate change impacts drought as we have seen in California. These droughts lead to the trees becoming vulnerable to attack from disease and beetle kill. As these occur—dying trees combined with drought—they are highly susceptible to fire.

If the forest lands have not been properly managed and you have a collection of ladder fuels on the forest floor, a fire ignited can become catastrophic as there has not been enough done to maintain a healthy forest with selective thinning. The number of trees that can grow per acre is due to the lack of treatment on the national forest lands, which in turn, is due to a lack of budget and a lack of personnel. The density varies according to the type of climate and terrain.

There are very clear examples. The forest service has done a very good job of demonstrating that forest lands that are treated for long-term sustainability and selected logging practices with the removal of the high-volume of biomass and ladder fuels on the forest floor have a lower temperature fire going across the forest floor, as opposed to areas left untreated resulting in a catastrophic fire destroying thousands of acres.

An additional problem that foresters understand is that when a catastrophic fire goes across forestlands, it kills everything that comes its way. It takes years to recover, but at the same time, an immediate impact occurs in the form of plasticine characteristic of soils that develops from catastrophic fire. It so happens that a catastrophic fire also kills the microecology of the soils within several inches and impacts the affected soil and root's quality in a mesh of debris that retains water and holds the soil together from times when it does rain.

A healthy forest will have the ability to hold more water being known as the least expensive way to treat water quality as it runs into the ground and percolates through the ground systems into the streams and rivers. Thus, water is cleaner and it costs less when it has been treated naturally.

Also, treated forests that will be able to resist catastrophic fires are better for the natural environment that includes healthier streams, healthier woodlands, and healthier forestlands as the habitat for all the animals that live there.

These are all the characteristics that come from healthy forests, and this is something we want to strive for. I believe I mentioned in an earlier chapter that we are going to have to look to some slightly modified forms of remedies under which we treat our national forests. We can provide sawmill material for the lumber mills and increase the number of local jobs on a long-term basis while we are providing means for long-term sustainable healthy forests in communities since there will be multiple streams of services and products needed, which includes more than just wood products or pulp and paper products that are going to be extracted from the waste streams.

Pulp and paper need a higher quality chip—a number one chip is what they often talk about. Sawlogs need to be of a certain dimension although currently, they have gotten smaller. Removing the understory heavy fuel loading will be a way to utilize small woody waste material that can be used in a biorefinery for biofuels and biochemicals including biochar for carbon storage in soils..

This is an additional opportunity in producing and developing new technologies that can provide a portion of non-fossil fuel to the transportation industry while simultaneously helping support healthy forests. In doing this, there is much work that has been done and studied to bring this forward even though there is an obvious lack of coordination and a structure in policy to facilitate all this.

Some years ago, I wrote a white paper for the Clinton White House that created a biomass coordinating committee where federal agencies could begin to address issues of national forestlands both on BLM national forests and of long-term forest health in a way that would produce solutions. The solutions drawn out were, however, not implemented at all. There was a cooperative memorandum between the Forest Service, the USDOE, the EPA, and BLM to create a joint plan. It did not mature into any action.

Another way for us to lower the thermostat and turn down the CO2 in the atmosphere is for people to do simple things that pile up into massive outcomes. That can amount to weatherization of their homes and having simple things like a smart thermostat. These products and services are available currently from a group called Nest or Eco bee, and even Honeywell has stepped up the market with a new smart thermostat.

Energy conservation through new digital equipment and weatherization programs by a local utility can lower demand. When you lower demand across the country, you will be able to offset some of the lower demands that will facilitate shutting down the coal plants, which they have already started doing. While lowering demand can take time, there is the chance to build renewable systems to replace the closing coal power plants. While the Trump administration has turned its nose up at climate change, which is of course stupid, there will be, nonetheless, other opportunities as we go through these kinds of new technologies that really can provide the same amount of power but does it in a way that would not pollute the environment. Fuel cells using natural gas can be a source of neighborhood modular systems that can function on demand for extra power.

We have opportunities for renewable energy power plants. The ones you are familiar with like large windmills and big solar arrays are well underway. I am introducing and had introduced a little while ago a geothermal/solar thermal hybrid power system.

In this system, which I will fully address in another chapter, geothermal heat of low or moderate temperature is used in the design to provide heating to the feed-water/heat transfer fluid from a concentrating solar power plant, which would then operate at a temperature that is much higher. An air-cooled power cycle from a Rankine engine is driven by the CSP plant with inlet conditions that are composed of superheated steam and a gross conversion of thermal-to-electric that occurs with an efficiency of around 15%. There is no emission attached to that process.

Today, there is one plant in Nevada that has proven this concept to be viable at a utility scale.

The Department of Energy has energy conservation programs for federal buildings across the country. States have implemented large-scale energy conservation programs to be retrofitted to all buildings. If you have not yet found out how efficient your home or the building where you work is, this can easily be done by getting in touch with your state's energy office or your local utility to find out about weatherization and conservation programs that can analyze your home and help you address the lowering of energy demand.

Another emerging technology that has been difficult to calculate is carbon sequestration. Carbon sequestration is absorbing CO2 out of the atmosphere, and the question is how we do that in a significant way that will help reduce the current calculation which exceeds 400 parts per million in the atmosphere, and it has been documented at the top of Mauna Loa in Hawaii.

Atmospheric carbon dioxide continued its rapid rise in 2019 with the average for May peaking at 414.7 parts per million at NOAA's Mauna Loa Atmospheric Baseline Observatory. Scientists from NOAA and Scripps Institution of Oceanography at the University of California San Diego announced recently.

We can address these issues with the large-scale growth of seaweed and algae in the oceans and various areas, especially in the dead zones. We are going to have to lower our use of fertilizers, pesticides, and insecticides across the globe. In conjunction with that, we are going to have to make sure that we understand that CO2 in the atmosphere is also the one getting stored in the oceans. The oceans, as a result of increased carbon dioxide levels, have increased their acidification levels as well, which has a massive impact on the plants and animals living in the oceans. Evidences of the problems in the ocean are clearly visible with the dying of the Great Barrier Reef off Australia.

We are going to have to address preserving large areas of native forests and old-growth areas while enhancing those on a long-term basis with monitoring and selective harvesting.

There is some carbon sequestration that is being pioneered at the Pacific Northwest Labs where they are sequestering CO2 in deep rock formations. Again, this is at a pilot scale, and some work has proven to be viable, but that is not manifested into a large-scale application that can be termed universally applicable.

We have to establish areas where cities make better plans for intra-city mass transit using low-emission transportation systems, fuel cell-powered buses, and electric mass transit. We also need to encourage communities to use hybrid automobiles and electric vehicles by providing adequate recharging systems throughout the communities.

Another form of remedy that we are going to have to explore is high-speed rail in the U.S. between major cities. The airplanes industry is a major polluter in the atmosphere, and things can be adjusted by lowering the demand for airplanes since I am sure people will not miss getting patted down if they can travel between some major hubs at some 250 miles an hour.

Those are some of the technologies and ways in which we can turn down the thermostat. Weatherization programs across the country for low-income people should be a special target for communities to help everybody benefit from the opportunities of conservation and energy savings.

Various items can be found at Home Depot as a network, Ace Hardware, and other stores as a means to seek easy ways to save money and energy.

Chapter 16: Paying for the Work

One of the other tough parts besides policies, technology, and political differences for fixing climate change is how to pay for it. There are a number of proposed solutions, some of which are favored by some folks while rejected by others, and so, it goes on with almost every other proposed remedy. There are always some real conflicts.

One of the proposed solutions includes cap and trade, which is a layman term for a regulatory program run by the government designed in a way to cap or limit the net emissions of greenhouse gases. Another approach is carbon tax which would have to be spread across the fossil fuel industry so that there is a tax on the emissions of fossil fuels from the coal industry, power industry, aviation industry, and auto industry. All of those would need to be included to pay the carbon tax.

In addition to carbon taxes, we would need to remove all of the tax credits that fossil fuel industries have been benefiting from. We need to remove the tax credits to create a more level-playing field for renewable technologies to be a little more price-competitive even though we are seeing substantial improvements in cost generation from wind and PV technologies.

This still does not mean that we will have an entirely level-playing field for the environmental benefit because the real cost to the environment also comes in the form of some catastrophic weather events that we see in the form of floods, torrential downpours, catastrophic fires, and mudslides occurring all across the planet. There is no strategy yet devised to price that into the increased emissions by fossil fuels and the way we use energy today, and that has to change.

There could be several approaches applied to fix that. One approach is a straight fossil carbon tax. This has been suggested but has not passed in any US state. Another is a cap and trade approach. This has been proposed in Oregon and elsewhere with no success. Washington state had a measure on the ballot for a carbon tax and it failed.

We will need to consider that the growth of alternative fueled cars means we need to address the lower fossil fuel tax revenues. Solutions will need to include a vehicle mileage tax so that the cars that no longer use the fuels can be taxed. Not only the fuel-driven cars but the hybrid and electric cars also need to be added to this tax since they are using the roads. It has to include maybe a vehicle mileage tax so that some of the costs of the roads and repairs that industry trucks or cars, even if they are hybrids and electrics as long as they use the roads and that we have to figure in a factor to mitigate the loss of sales revenues from gasoline sales and diesel fuel sales, has to play into the calculations.

We may need a special tax credit program for renewable energy and utility systems. There is an example of a program many years ago that became very successful for moving the construction of renewable facilities forward called Section 1603. We might want to reexamine how successful that program was and use it as a model to go forward.

Homeowner programs for weatherization and retrofitting of new energy-efficient windows and increased insulation have been a part of a national energy weatherization program. Following through what was started earlier and has been ongoing should be continued and strengthened. Weatherization of homes for people with low income and seniors needs to be added into the mix of everything that is going on, and it's not just for homeowners but for business owners and commercial buildings as well.

There are multiple programs and technologies for increasing the efficiency in lighting, heating, and cooling of buildings, and this is more popularly known as DSM or demand-side management. There are ways in which, depending on the amount of load put on a system in communities, the utility can manage that load with demand-side management control technology. New wireless remote communication systems that can communicate with a utility can allow a utility to manage the energy demand; for example, a hot water heater, to not turn on when the demand is high but turn on when the demand is lower.

In addition to that, there could be a new approach by renaming the Rural Electrification Administration (REA) to Rural Renewable Electrification Administration (RREA) through which we use the existing structure that helped in expanding the transmission and distribution infrastructure into expanded rural areas that are not now accessed. Areas in Eastern Oregon and Eastern Nevada are two examples where the T&D system does not fully extend to tap into renewable systems. They can facilitate connecting new renewable energy systems that would not be convenient or economically viable for rural utilities as well as the interconnection for renewable systems that might, by the nature of their generation, are in more isolated areas from where the power is used.

This would come into play for wind, solar, geothermal, and geothermal solar systems.

There is another approach as to how many new community systems may be developed that can be nonprofit and use funding from the likes of the new energy fund that has been touted by Bill Gates and Warren Buffet since they are planning to finance billions in renewable energy.

If we have communities that create nonprofit renewable energy entities to implement the green energy strategies by using those funds in the form of grants or low-interest loans on a long-term basis to facilitate the cost of the phasing of renewable energy systems, then it can create real solutions for our environment.

Another way to address the funding of these programs is by having an increase in the estate tax or inheritance tax. Inheritance taxes by nature have been at historic lows, and there is certainly adequate room and funds available for using some of these as set-asides from large estates. Billion-dollar estates should be the prime source of targets in the estate planning.

In addition to this, a small surcharge on all energy sales in communities throughout the country at a few mills per kilowatt could also create a renewable energy fund to facilitate the implementation of across-the-board renewable energy—the necessary interconnection requirements costs.

Chapter 17: Problems with Large Corporations and Capital Concentrations in Implementing Large-Scale Renewable Energy Programs Across Countries and the Globe

The French economist, Thomas Piketty, in his book *Capitalism in the 21st Century* discusses the problems with the disparity of wealth. The concentration of wealth and capital for the benefit of a few diminishes the economic opportunities to the larger populace.

There are real strengths to community solar projects, community-scale cogeneration, and windmill farms where the wind towers are spread over a variety of farms. These choices support the more local economic model that can help keep resources in a community, which ultimately adds strength to that community.

A windmill on a farm that receives lease payments can make a difference for the family on that farm by sustaining their farming activities without them needing to sacrifice time for a job in town to support the needs of a family farm. Of course, the lease payments would need to be made to the owners of the land whom it is proposed is the farming family. It may be necessary to create a new class of ownership for historical family farms to prevent corporate abuse on the benefit of a prioritized windmill on a family farm.

A utility-owned windmill becomes enormously valuable as the revenue offsets mean the family farm has a greater opportunity to succeed with a sustainable farm even with organic farming techniques being implemented on a long-term basis. There are programs currently in place at the Department of Agriculture to facilitate sustainable organic farming. The availability of solar and wind towers on a family farm receives lease payments from the utility for that facility being installed. As I say, it makes all the difference to people when they are able to stay on their farms even with the vagaries of weather and water shortages impacting their successes in the farming industry.

Community solar means that there are cost efficiencies for a group buying the solar projects as opposed to just a single homeowner putting a system on his home. This however comes without the higher net cost that is usually allocated to the smaller system in one home. It allows lower-income families to participate in the benefit of renewable energy by efficiently sharing the cost of the solar PV system with the whole community for mutual benefit.

There are cost-sharing benefits that accrue to everybody in the community; therefore, the entire community has a greater economic benefit. Doing this will create an economic advantage for communities on a long-term basis.

The current structure allows large corporations to take advantage of the solar credits, and by having the kinds of financial muscles that they do have, they reap real significant benefits. These benefits, however, are accruing to the corporation, and while that is helping the environment and giving the corporation a good environmental profile, it is still an internal rate of return calculated by corporations for that corporation's benefit, and not for the benefit of larger communities or to the economy. One should keep in mind that renewable energy is for the advancement of the economy and community, which might be halted by the corporations using it to their advantage.

There are also large capital investments that take the kind of financial strength that only corporations have, but a community that pulls itself together and has the community resource of bonding available or the means of creating a nonprofit company available can also have a significant benefit to the community on a long-term basis.

For example, if the renewable energy facility has been capitalized and paid off, be it wind or solar, then those higher costs would have been compensated as well. The cost of energy drops consequently, and that lower cost of energy accrue to everybody in the community, not just the large corporations.

This is one of the reasons why it is important for us to look at the community-scale facilities that can, in the process of having paid the capital cost of the facility and once the capital cost is recovered, lower the cost of energy for the community. This ultimately can be a significant cost saving that spread through the community adding economic benefits throughout the populace.

It would have been my particular choice to build such a public facility that would be community-based so as to benefit the whole cities and towns, but unfortunately, that's not how the financing of renewables is usually structured.

There is a biomass cogeneration power plant that I established in Medford, Oregon called Biomass One, LP. That plant has now paid its debt; it has been running for 30 years, so now on the private side, the energy revenues are accrued to the owners. There has been a drop in the overall cost, which has gone to the utility paying a lower cost of the energy, but at the same time, there is a significant benefit to the community with other local jobs that have been created and sustained for

30 years, and along with the lower emissions in the community that made significant strides in cleaning up the air.

I do see a future in the renewable energy systems that are community-based like solar, geothermal, and wind. If we can have them under community ownership and public ownership, that opportunity of a lower cost of energy accrues to the community and self-support the long-term success of the economics of the local community.

This is a real opportunity for communities to have a long-term impact on lowering their utility costs and having the benefit accrue to the community by lowering utilities' costs in the communities.

Another issue is that large capital investments are for the benefit of the company and not for the community in a long-term.

Another illustration of large concentrations is nuclear plants. A nuclear plant that has been typically 1,000 megawatts will run in billions of dollars in one facility. What if the facility is unneeded with the benefits of building design for energy conservation, considerably lowering the use and demand by housing, is spread amongst the renewable systems in various locations?

This means that the economic invested capital goes into those communities for sustainable systems. In addition, when you have a solar thermal system or a geothermal system, it is an opportunity for district heating so that the variability of business demands for thermal demand is met on a local basis.

Greenhouses add to the variability of the community to support itself during wintertime to grow local organic crops and provide long-term employment.

There are other illustrations of these kinds of community models whether it is BPA, Clark Public Utilities, Eugene Water & Electric Board, or Tennessee Valley Authority. The power grid can integrate and deliver renewables with that kind of public-supported system.

This also means that there is the opportunity to introduce micro grids in development that can also access the larger grid but also make it independent.

Chapter 18: Benefits to Communities in Developing and Owning Their Own Renewable Facilities

In the previous chapter, I made a reference to the value of a community-owned system. Here, I am going to try and delineate additional benefits or further explain that there are multiple benefits to community renewable energy projects that accrue to the community, supporting the investments in such systems that support the local economy and add value over the long term to the community by lowering energy costs on the long term.

In addition, community jobs with a support system for the success of the community's economy can be accredited to these kinds of renewable energy facilities. There is also the long-term investment in the community infrastructure providing again long-term sustainable benefits and a lower cost of energy which strengthen the local economy and extend the value of the resources locally.

Lower and stable energy costs accrue to the community over the long term. There are more opportunities for members of the community to have local jobs at family wages—a stronger economy means a stronger education system and a stronger education system means that the STEM programs science, technology, engineering, and mathematics will have greater value within the community. All of this will provide immense strength to the educational system of the society supporting additional diversity in jobs at family wages and beyond.

In any well-designed renewable energy system that happens to have a biomass plant, a solar thermal facility, or a geothermal power facility, there will be opportunities for greenhouses. In these greenhouses, of course, depending on the climate, any number of crops can be grown.

Additionally, not only greenhouses but aquaponics/hydroponics become opportunities for communities to have the additional infrastructure to support their local economy. This can also be a benefit to the district heating opportunities from biomass, solar thermal, or geothermal energies. There are opportunities for additional jobs and industries created from heat processing and the value of that processed heat.

One of the things that seem to be little understood is mostly in the areas of district heating. Taking geothermal as an example, you need processed heat to heat the greenhouses in the winter. The advantage of this kind of heating is that you do not really need a cost for heating a greenhouse in the summer, and at the same time, if you do not need the heat for the generation of processed heat, then you can always reinject it into the ground. Thus, heat is not generated via some expensive process using natural gas and then expended into the ground. It is rather generated via reinjecting it into the ground only to be recirculated later on.

This is one of the areas that are widely overlooked and there are opportunities for district heating since there are tremendous numbers of industries that need processed heat. It can be fish farming, growing shrimp, and for use in other crops.

Examples that are in support of district heating are largely found in Europe. Klamath Falls has geothermal district heating in Oregon. Reykjavik, Iceland is probably the premier example of the high worth of district heating by geothermal means.

In addition to district heating technologies and opportunities, there are also well-established models in the public sector like Bonneville Power Administration and Clark Public Utilities as a distribution facility for the BPA system. You have EWEB and Eugene Water and Electric Board, which is again, a public utility down in Eugene that distributes energy to the community. In the Midwest, it is the TVA or Tennessee Valley Administration providing power to the grid in that environment.

The new technologies that are emerging with renewables also are creating opportunities for microgrids. Microgrids are something I talked about, and I mentioned these emerging integrated facilities at a smaller scale at the Verge conference.

One of the things here is not just the community opportunity but a significant economic opportunity benefits for the development of industries that export these microgrid technologies to developing countries in Africa and in Asia where a whole full-blown power system, power poles, and large power-producing plants are not in place. An example of the value of a micro grid is now being built in the Bahama by Rocky Mtn. Institute where the power is feeding a school and a hospital since the old power system has not been rebuilt. The smaller systems can fill the void where no systems exist or have been destroyed.

This is a tremendous opportunity for foreign affairs that are state department supported. AID-facilitated export of micro grid technologies across the developing world on a basis of renewable systems is established so that communities can have some light and power to run refrigeration, light schools, and little villages that add a tremendous opportunity for education in order to implement better healthcare and run rural pumps for water from wells. This could be a tremendous asset in our foreign aid program—something that we export, which has tremendous value for people in emerging countries.

That is something that will give the communities long-term benefits while we give them the training to maintain the facilities that are not so high-tech or dangerous. In this manner, they can really benefit significantly and create a more sustainable environment politically, economically, and culturally by respecting the culture of the local communities once they are adapted to the modernization. This, however, is without the full-blown infrastructure that we may not find economically viable to put in place for them.

Using the new technologies like LED means that lighting will come at a significantly reduced cost. Additionally, it does not necessarily have to be 110 volts or 220 volts. Some of the lighting systems run on 18 volts, which is very adaptable to PV with battery backup. The same is the case with adapting from wind through the generators and an inverter, so there is an opportunity of adapting the new renewable technology to a less energy-intensive usage but still providing the necessary amounts of power to run small laptops and lighting in villages, in schools, hospitals, health clinics, and these kinds of public facilities in the emerging world.

There are over 1.3 billion people who are living without electricity and energy and the benefits from it.

Refrigeration is one of those items that can preserve food and critical medicines. Without refrigeration benefits, these become critical issues. I am remembering a friend from years ago who built a six-feet-long parabolic trough collector attached to a refrigerator that used the heat from the collector to drive a refrigerator compressor to cool a refrigerator. This type of tool would be most useful in warmer climates.

At this point, one of the things that I have thought about is the steps taken to move the discussions toward the use of renewable energy as a pathway to peace in the Middle East and across the areas of conflict throughout the world.

The availability and adaptability of renewable energy across a broad spectrum of the geographical areas occupied by human beings mean it is an opportunity to minimize the territorial issues of availability in resources and provide power, opportunities for education, lighting, and all of the other attendant economic benefits that can come from creating a sustainable, renewable electric system in rural communities.

Something I wrote up a number of years ago is a way to address what could be some baby steps toward pathways to peace based on renewable energy.

Chapter 19: Necessary Steps to Take

In order to make the progress we need and spare our future generations from tragic weather events, world-impacting disruptions where millions of people will be displaced, and the trauma of millions and millions of refugees from elevated ocean levels, we have to have an effective plan of action to combat for climate change where the forces opposing this are seeing the end of the fossil era.

This does not mean the complete end of the gas and oil industry; they just are going to have to live with serious curtailments.

How are we going to get there given the misstatements and misunderstanding that some numbers of our politicians have? For one, you are going to have to take an active role in examining their positions and statements that they take about climate change, economic development, support for renewable energy, and a serious understanding of what underlies the causation.

For example, the Trump head of the EPA is saying there is still some questions about climate change and he, I believe, discounted that carbon dioxide is a greenhouse gas. Somebody like that, in my opinion, is giving evidence of prima facie ignorance and has no business serving the public. He should be relegated back to his house, wherever he is from, and told to sit down and shut up because he does not know what he is talking about, has no idea about the impact, has no solutions, and is contributing to the problems and the demise of stability across the globe.

You are going to have to be a person who learns to think critically and ask questions about policy and watch what people are doing in formulating appropriate legislation. From some of the areas in this book, you will learn what has worked and where it has worked and how to understand where you can get this information.

It is going beyond the divestment mentality of simply selling your oil stock in your portfolio. You have to go further. That is where creating community-funded systems that provide a return to the community and its members in that community project stand to benefit. The internet is inventing new solutions for funding renewables—look there.

Chapter 20: More About the Necessary Steps to Take

In order to make the progress we need and to spare our future generations from tragic weather events and world-impacting disruptions to millions of people, we must have an effective plan of action to combat the forces that oppose the end of the fossil era. This does not mean the end of the gas and oil industry; they just have to live with some serious constraints.

I do think that there are technologies we have yet to see and emerge that can use that type of carbon-rich feedstock to create very economical structures for people to live in and that can be molded and formed in panels that can interlock and be locked together that create habitable facilities and housing on a broad scale. We have not tapped that resource for those kinds of applications, plus the recycling of waste plastic into formed material.

One of the things I saw recently was using 3D printing from a recycled plastic feed—sort of a rope of recycled plastic fed into the 3D printer that can create usable objects which designers can design and do a computer-aided design to produce different functional materials.

Additionally, something I saw recently was a 3D printer printing out a concrete house. There is an example of where we will be going and developing if we have the resource like mixing together a lightweight foam concrete with plastic that can be very resilient to the weather resulting in long-term housing that does not degrade in the environment for a number of years and can provide housing for multiple generations. That is something we could strive for.

That is just illustrative of some of the steps we need to take and be creative about and at the same time recognizing that the oil and gas industry has a different role to play.

In further examining, the steps to take for going forward is the political climate. What I am going to ask you to consider, readers, is your elected representatives and what you may want to do like making sure you review and look for position statements on a variety of topics like the environment, renewable energy, and organic foods. If there is no position statement by a candidate, you may want to look at other candidates regardless of the party.

The other question is that are they prepared to formulate appropriate legislation? Would they look at what has worked and where it has worked, and if so, would they support it? If not, why not? You must ask questions to people, and you will have to go beyond the approach that simply says divest of the oil industry.

One of the things that needs to be done is that people need to come to the point of deciding to invest in newly emerging technologies. Community-funded systems, which we begin to see happening through the internet, is another way to fund renewables, and this is emerging as we review the opportunities for alternative funding.

Chapter 21: Obstacles to Progress We Must Address

The road ahead is a hard one but it's important that we take it because it provides tremendous opportunities for the future that people will benefit from for generations.

Reading through this book, you have learned about technologies that are currently available, can be implemented, are being implemented, and available to your community as well as technologies that are available to go across the globe, and we can agree to fix these problems with the right choices.

This is an opportunity for exporting safe technology that benefits people across the globe—to live healthier, have an affordable infrastructure for electricity and for thermal process heat, and not give up much in the way of what modern amenities and inventions have brought us. It does address the issues related to the over usage of the limited resources.

There are obstacles that we are going to have to address in the utilization of biomass energy and doing that in a way that supports the long-term sustainable healthy forestlands we have. We are going to need our forestlands to store carbon dioxide. Trees breathe in carbon dioxide, store the carbon in their lignocellulosic structure, and emit oxygen.

We are going to have the opportunity to use the biomass coordinating committee that was established during the Clinton administration. Information on this approach is available about the resources in the western United States that was done by the Western Governors Association and their biomass task force. That is available online.

Another is that there is a study that was done by the Oakridge National Laboratories, and that is a biomass report of a billion-ton waste stream. What was determined by the national laboratory was that there are a billion tons of wood waste already accumulated to be disposed of across the United States. If we take a few extra steps and take that wood waste stream and put it to being processed for biofuels, we can substantially reduce the impact of fossil fuels.

A cellulosic process and research that is being funded for biofuels conversion technologies have been going on for a couple of decades. These technologies, as you will have read about in the book, mean that the cellulose can be converted into sugars, those sugars can be converted into ethanol, and the ethanol can be converted into biodiesel, which can then be converted into aviation fuel.

There is another process that would involve creating a syngas, and in converting the cellulose/wood waste to syngas, those dirtier feedstocks would be able to have their contaminants mitigated by the high heat of the syngas creation. That syngas can be cooled and processed through an anaerobic process that converts it into ethanol. Again, the ethanol can be converted into biofuels.

Biofuels conversion technology has been going on for a number of years, and you will be able to find that information in searching the internet.

CO2 sequestration is another problem that has to be addressed. There is some success in sequestering CO2 in rock formations underground, but that has only been done at a small pilot-scale level by the research scientists at Pacific Northwest Laboratories. CO2 can be stored in the soil with biochar also.

There are some opportunities that are not properly or well-funded for carbon sequestration, which is growing algae and seaweed in the oceans on a large scale. Another problem we are going to have to deal with is the acidification of the oceans.

Excess CO2 is being stored in the oceans. What this does is create carbonic acid and change the pH. That change, while small, has a profound impact on the ecology of the ocean, and the balance in the ocean is changing so you now have dead zones, large die-offs that have been recorded, and starfish and other small invertebrates that have been dying.

You see that some of the sea life—the large mammals that feed on these smaller systems and the fish they feed on are turning up to be starved because of the lack of food in the ocean brought by this impact on the ocean's chemistry.

This has to be addressed immediately, and yet, it is not receiving the attention and the funding it deserves.

Recent newspaper reports show that Australia's Great Barrier Reef has about two-thirds of it bleached out, and this is an extremely vital environment for ocean life and its health, especially those that are dying. This is an example of the canary in the coal mine.

Sealife is something that is critical to the full ability of the earth to sustain life, especially the complex life that we live.

These things have to be addressed. Look at making a questionnaire. It is your responsibility as a member of society to address people who would represent you to make sure that they have an understanding of science. If they deny science in any way, they are not acceptable to be representatives for the public welfare.

Renewable energy and renewable technologies, I think, are easily one of the most opportune technologies to present to the developing world with solutions for poverty, and we owe it to others on our planet to implement solutions in sustainability with renewable energy technologies that we export to other parts of the globe so they can move forward in creating sustainable environments and rehabilitating the desertification of forests across vast areas of the globe.

In summation, the technologies are here now to move to a cleaner energy-efficient world. We need the political will, driven overall by the people, to drive the politicians to enact the legal changes needed to affect the best outcome. Modification to existing rules in the Uniform Building Code (UBC) and banking rules to support higher energy efficiency standards are examples of how we get the improvements we need. Be attentive to supporting leaders who are proactive for the right solutions.

Interview

This interview is with my friend Don Bain, a wind energy pioneer with Aero Power.

Don worked for the Oregon Department of Energy for many years as the wind expert. Upon leaving the ODOE, he worked to establish a major 50-Megawatt wind farm called Combine Hills in Eastern Washington.

MARC: Hi, Don. I'm really glad you came over so we could have this chat about wind since I don't know the ins and outs of wind the way I do biomass or the solar-thermal hybrids and geothermal, but you've been doing wind quite a while. How many years is it you started doing research with the Oregon Department of Energy? Why don't we talk about a little of your history and how you got into the wind and you did some of the early projects?

DON: Well, actually my history with wind goes a long way back before my start at Oregon Department of Energy in 1979. It started back in the mid-seventies at Lockheed, a California Company in Burbank, where I was working on the flight line, and in the hangar next door was a model wind turbine blade on a stress testing fixture, and I got interested in that. At the same time, I was taking an economics independent study course, so I decided to study alternative energy and focused on wind. In the process of that, I interviewed the managing guy for that wind turbine blade test and sent him a copy of my paper. After the class was done, I went back to my toolbox, didn't think a lot of it, continued my college courses, and then wound up getting hired over into his group in mid-74 to 1975 when the Energy Research and Development Administration awarded a large contract called wind energy mission analysis to Lockheed to study the future of wind energy for the next 20 years and when he had somebody disappear from his team in the interim. I knew something about wind, and I got drafted into it, which was a fantastic experience. GE was awarded the same contract, and we weren't allowed to talk to each other. We looked basically at the future of the technology, the costs, the application scenarios, the issues—all sorts of things for the 20-year forecast period all across the US. So, that was really the start of it. When all of that was over, Lockheed attempted to do a little bit more but didn't get any more federal contracts. A group of us guys did some moonlighting and did a very similar, smaller-scoped study for the California Energy Commission that was used by Matt Ginnesar and Jim Lerner to justify budget requests for setting up the State of California wind energy program, and one of the big elements of that was the wind resource assessment, which is how Rich Simon and a number of the early meteorologists got jobs to go out and figure out where the windy spots in California are, and they published a map and that mushroomed into the developments in California with tax credits for development in Altamont Pass, Gorgonio Pass, and Tehachapi. The other thing that happened later on was in 1978. PURPA was passed. This was right before I went to Oregon in 1979, so I was working on the behavior of the American Wind Energy Association to lobby FERK about the PURPA rules. So, I was doing that and then consequently went on a tour with FERC to present in five cities, I think, around the US with the PURPA rules, how they worked, and the whole business. Then, I presented that at the American Wind Energy Association annual national conference that here's now a way you can sell and you're not regulated as utilities, there are standard contracts, there is avoided cost pricing, there is a right to interconnect, and all these other important features.

MARC: Yeah, a lot happened with that PURPA legislation.

DON: Yes. Without that, it wouldn't have been possible to get an industry launched because you couldn't independently do projects you couldn't finance; utilities were protecting their monopolies.

MARC: And that was one of the drawbacks that people don't realize because the utility industry is being heavily regulated. They were opposed to the PURPA legislation.

DON: Pretty much.

MARC: And they did their utmost over the subsequent years to redefine the avoided cost formulas which is one of the things that impacted California's biomass projects when they did an SO4 contract, and they made it only a 10-year term, and after the 10-year term, the renegotiated rate was going to be based on natural gas which completely undermined those projects rather than longer term contracts. Has that changed?

DON: Well, it evolved over time. California took off with the PURPA legislation and final rules in 1980 with its own processes which set up the standard offer contracts. Oregon, after I got there in 1979, we were going into a legislative session and so I was working all the PURPA-related issues for the Oregon Department of Energy both in the legislature and at the Oregon PUC, and I had strongly recommended that because the utilities were trying to repeal PURPA all the way to the Supreme Court, there were a couple of cases where they attempted to just get rid of it wholesale. That failed, but because of this ongoing uncertainty and the time it takes, Oregon passed what we call the mini-PURPA which took the core pieces of the federal rules and put them into state statute so that we had a stable environment that could roll forward with that. And of course, the utilities, as you said, had quite a bit of interest in how you define avoided cost and the calculations and all kinds of things and raised all sorts of issues about interconnecting with their system and safety and so on, and that, in turn, led to discussions about wheeling. The PURPA legislation at the federal level addressed that but only indirectly and it provided only a framework for how, if you got wheeling, then you could arrive on a utility's doorstep in a different service district than where the project is, and you would be entitled to an avoided cost pricing at that doorstep. However, how you got there was still not really covered under the federal PURPA rules, so wheeling became a big issue for quite a long time. Then, as the PURPA avoided cost issue evolved, utilities came around to say, okay, if we're going to have to do this, how do we keep the cost at the minimum to the ratepayers? We don't like these avoided cost calculations; they require us to prognosticate the future; they're difficult to predict and so on and so forth; let's just do an auction. And so, they went into competitive bidding for contracts and the net result of that after a number of years of this adoption of competitive bidding, which was sanctioned by the states, was that competitive bidding really set what the avoided cost was rather than some calculated number and model which then meant that every developer couldn't go to, let's say a table like under the standard offer contracts, and say alright, if you did this, here's the price.

MARC: Yeah.

DON: Now you could go out and figure out on the basis of your price whether a project was viable. Competitive bidding threw all of that risk back on the developer, so they had to now imagine what the price is likely to be, what they think is viable, what they think the competition can do it for, what this project and that project is going to cost, and now that meant that all of the development capital was totally at risk because you couldn't really count on what the actual final price would be nor could you count on a contract. All you could do then was assemble the most cost-effective project you think you could find, throw it into the competitive bidding hopper, and hope that you're going to win the PPA, but if you couldn't win the power sales agreement, then all of that was basically wasted money—millions of dollars and years' worth of effort. So, it evolved out of a fairly price-stable and known marketplace into one that created opportunity, but on the other hand, an awful lot of risk and financial demands on the developers. So, as that happened, that meant that developers were taking more risks with more and more money. That brought in the financial community into the qualifying facility marketplace and of course, once that happened along with the bidding, money was looking for any kind of opportunity to invest in energy. Utilities opened up competitive bidding to not only renewable energy facilities, which qualified under PURPA, but they opened it up to virtually anything. So, they'd be saying okay, we have a need for X-amount of megawatt-hours or megawatts this year—come one, come all, whether you're a qualifying facility or not. So, that was a further evolutionary step in this whole competitive environment. That, in turn, further evolved to where a developer with a renewable energy facility could find himself competing against the utility's own facility to build, let's say, a natural gas-fired power plant in the same auction.

MARC: And that is still occurring, I think, in some places.

DON: It's alive and well.

MARC: Yeah. Makes it a little difficult for an independent power producer.

DON: Yeah, and the whole idea is that there is a Chinese wall within the utility.

MARC: What does that mean for people who don't know the subject matter?

DON: Well, the utility is supposed to operate in the interest of ratepayers. On the other hand, organizationally, legally, and financially, it's operating in the interest of its shareholders. Since the utility has a monopoly, then there is a conflict of interest here sometimes between what's best for the shareholders and what's best for the rate payers, and if this is a monopoly, there's no competition to drive that price down for the ratepayers. There is only one possible seller, and that's your utility. So, the proxy for that was that utility commissions would tell the utility you have to set up a barrier of communication and organization within your company so that you don't have this bleed of conflict of interest from one side to the other between what's in the interest of the ratepayer and what's in the interest of the shareholder. Where those diverge, they can't cross-contaminate, and so when you go out and buy new energy resources to serve ratepayers, it has to be strictly evaluated on the basis of what's good for the ratepayers and not on what's best for the shareholders. So, in order to accomplish that, this artificial wall within the organization to keep this conflict of interest from occurring has to be created. So, where the people who are putting together the auction process and evaluating the proposals and making those decisions and all that, they're not supposed to think about what's the impact on shareholders. They're supposed to evaluate whether the utility self-provided alternative is better or worse than what the outside market can provide without any bias, and so on; hence, the word "Chinese wall."

MARC: Okay.

DON: It's this separating wall between these different goals and objectives and how they are administered within the same company.

MARC: It sounds like the fox watching the hen house—a little bit of that . . .

DON: Well . . .

MARC: . . . even though they say they're a separate entity.

DON: Yeah. And we have a Chinese wall that is federally administered between transmission and generation because the transmission has become more or less a common carrier, which is open to all and which the utility can't say no to except for technical constraints or reliability constraints, and if the transmission system needs upgrading, then the requester has to pay for that upgrade. So, that's difficult than the interests of generation, which is not a common carrier. It's paid for by the ratepayers to service themselves. So, again, that's a Chinese wall between transmission and generation which has existed now for quite some time in the way that it's been imposed by FERK. So, these organizations we're talking about here, they've got multiple Chinese walls within them depending on the function or the process at hand.

MARC: Right. Now, what if we talk about what it takes, you know. So that's sort of the general structure. The PURPA was passed in '78. It got refined, it got attacked by the utilities and went to the Supreme Court, and it's been in force now since all of that has transpired for all these years. As I recall, the formula for avoided cost has really changed from what it originally was written and has gotten looser and more favorable to the regulated utility's application of the least cost resource. And in fact, I think today's wind is one of the least-cost resources that's available.

DON: Last year, the wind was the leading source of new capacity additions in the US, but this whole business, on the one hand where you had the advent of competitive bidding come, it sort of becomes the proxy for what the avoided cost is because remember the definition of avoided cost to the original PURPA, it was here as what the price would be if the utility were to do it by itself to supply new generation.

MARC: Yeah, and I had one of the very earliest PURPA contracts, and it was an avoided cost formula determined by the PUC based on the latest generation facility, which was Trojan Nuclear Plant in Oregon, and the utility was just screaming mad that they were going to have to pay the avoided cost pricing the PUC had formulated initially. Then, they argued that price down which the PUC agreed to in part because it was going to be a 30-year contract for power purchase, and they shoved a great deal of the revenue off from the early years to the back years, which actually hurt the project itself in the early years. Yeah, I saw that change and evolve as the industry matured and the utilities got smarter about these PURPA contracts.

DON: Well . . . but what you're experiencing there is this conflict of interest. The Trojan Nuclear Plant or the utility facility is highly capital intensive, and the return to shareholders is based upon invested capital in the utility facility, so from the shareholder's perspective, the company management is doing its job—it's trying to protect that interest and trying to gather more capital investments under its hat and make more regulated returns to that investment. The utility doesn't make a dime off of a power purchase agreement with an outside entity, so all of that capital investment is borne by the developer and the private financers out in the marketplace. You saw that, so of course, the utilities were arguing against these PURPA contracts because they displaced the opportunity for shareholders to have a growing set of dividends based upon the capital assets of the company. But going back to what we were talking about—the auctions—as those auctions matured and the auction pricing became the proxy for determining avoided costs over in another sphere, the whole concept of least cost plans was evolving and also coming into the utility industry. So, these two concepts merged because the concept was how do we supply ratepayers at the least cost? Well, now we'll do a plan; we'll examine different alternatives; we'll examine all these other things, so what might have looked like before as an avoided cost determination has now evolved into this broader scope plan, the least cost energy plan, which would then sort of formulate a future framework for whether or not the utility needs more resources and what types of resources look to be the cheapest—whether it's conservation or demand management or combining control areas, distributed generation, or capital intensive, let's say, baseload plants or as-available renewable energy plants on the margin or whatever—these things would be sort of categorically evaluated, and the least cost plan would then say, oh yeah, this looks like the best way to assemble this future. Now, of course, all was done within a different policy framework, but that policy framework was also related to the same policy framework that PURPA had, which was left to their own devices. We're not getting enough renewable energy here; we're not getting participation by the market in supplying these resources; we're not getting the innovation and the new industries and everything else under this monopsony model, so we need to go out into the market. PURPA was a vehicle for that, and of course, the favorite son in that marketplace because what was being systematically neglected was renewable energy; hence, PURPA focused on that. So, that sort of policy framework carried through to a lot of the policies that had to be evaluated in the least cost plants. So, the least cost plans would then say okay, we got to look at these, and the PUCs as a matter of public policy would say alright, you have to evaluate all of these different categories of resources to try and get a good feel for what's the best option here given some sort of scenario choice. What is the best way to proceed? It might be they decide oh, we can do all of this through conservation—we don't need a new generation, and that has been a bugaboo for the renewable energy industry, too. A large contingent of interests out there has long argued that we don't need any new generation. We can just take care of all of our growing needs through increasing conservation in the existing stock and therefore, free up whatever we need and make a few technical adjustments around the edges, but we can do it. So, you had this merging of least cost planning and auctions, and then the auctions, you see, then became the servant of the least cost plan. So, the least cost plan policies in that jurisdiction might say okay, it's our policy to go after renewables or it's our policy to XYZ, then as a result of that, the plans would focus on that. If the plan was accepted by the utility commission, that was a blessing of that sort of future at that level, and then the utility could move forward in that direction with some degree of comfort that they wouldn't get hammered when committed later on. Now, of course, the utility couldn't go in and ask for preapproval of a contract—something that doesn't exist yet—so how do they get from here to there? Well, they employ this auction, and they say okay, we tried our best. Here's the auction. It delivered up all these possibilities and here's the price.

MARC: Mm-hm.

DON: We did our best, you know, and then that added another degree of mitigation of the risks involved in making the resource commitment.

MARC: And while that's going on, there are technical developments going on in Europe that are advancing the technology for wind and energy production. It's not in a vacuum, you know, just happening in the US. I'm trying to remember how far back in the 80s that Bonneville Power did acquire a model large-scale wind turbine and large blades—I believe it was built by Boeing.

DON: Right.

MARC: And set up on the Columbia Gorge, and it sat and operated for a couple of years before it was decommissioned, sold cheaply at auction.

DON: Well, that was the original Mod-2 units from Boeing. There were three of them built up there near Goldendale, and they were built in a deliberate triangular arrangement so that they could investigate lakes from them, and as far as I know, that was the first time that the term "wind farm" was ever used, and that was in a speech, I believe, by the then current administrator, BPA, inaugurating the facility. And that did run. I mean, those were experimental wind turbines. Under the US DOE program which went from the Mod-O, Modo-OA, Mod-1, Mod-2, and Mod-3, and so on wind turbines. The objective of Lou Devone who ran that program was—let's push the envelope as far as we can technologically to learn what we can and let's develop the engineering, both aerodynamic and structural dynamic engineering so that we can produce models, have input, and know the right formulas and the right kind of things to pay attention to in these turbines. How do they react to off-axis wind flow, wind shear, or sudden changes in direction, to turbulence and things like blade flap with different ways of building them? I mean all of these things were questions.

MARC: Yeah, the scale was so much larger then. There were no pieces of equipment that were precedent.

DON: None. Well, I shouldn't say "none" completely. I mean, there were a few one-off very unique wind turbines that were developed, mainly out of Europe . . .

MARC: Yeah.

DON: . . .where people kind of on the fly, on the cuff, or whatever, put together a large turbine, and it might have been a university or it might have been some company where you have like Palmer-Putnam did in the US in World War II—a big interest in this and put money into it, private capital, and just said we're going to do it, and they just started doing it, but those events were virtually all one off units.

MARC: Mm-hm.

DON: They weren't the product of a complete system of understanding the science and the technology and trying to pull it all together like there was at the US DOE, but that didn't mean that the Europeans didn't make a lot of the early wind turbines. I mean, some of those turbines were made by companies that made farm machinery, and we call that sort of the Dutch model . . .

MARC: Right.

DON: . . . which were . . . the way that they handled things that broke was just add more steel, so they were heavy and in certain ways, were overbuilt, but they were rugged. They knew how to do that. Whereas in the US, the whole idea is championed by [inaudible] the feds evolved into, okay, we need to make these lighter. We need to make these with a minimum amount of materials, and if we just apply smart engineering, we can make them lighter and those ways advance the technology. With advanced technology, they'll be more cost-effective, and so on. So, there are these two competing sort of ideologies being pushed forward in the marketplace with the feds sort of saying okay, this is the way of the future, and so the money poured into that—the marketplace out there through PURPA and these auctions saying hey, we need wind turbines that are going to last 10 years and ultimately, now 20 years they're committing to the life of. They've got to be reliable and they've got to be cost predictable and so on for this long period—what works and the marketplace and the capital behind this. I mean, totally indifferent to whether this was the one built like a Sherman tank or one built like a hot air balloon. I mean, it didn't matter. What mattered was, was this commercially viable? What was going on there? So eventually the models, those two philosophies—they both cross-pollinated quite a bit, particularly in the aerodynamics and the blades and in understanding the drive trains and the structural dynamics, and a lot of that went into some early US company endeavors to go into the wind turbine business, but it also fed more mature and fast-evolving European set of models of wind turbines. They had basically a leg up with experience, so I guess you had to name it at the end of the day. It was sort of the European Model 1.

MARC: Mm-hm.

DON: In the marketplace, the federal program of lighter, higher tech got pulled back into basic engineering as sort of the marketplace settled all of these questions because now that the models were there and there was understanding of the technologies involved, they could be pretty well designed. That was probably, you know, in retrospect, a wise move on their part even though it was greeted with a lot of hand-wringing, the federal budget refocusing on basic engineering, out-of-marketplace development, out of applications, and out of all sorts of things that have evolved into, so now it's a completely refocused kind of program than what it used to be, and the technology in the companies involved are pretty much on their own.

MARC: Yeah. We have Vestas here in Portland with the headquarters here, based out of Belgium?

DON: Denmark.

MARC: Denmark. And Iberdrola has offices here in Portland?

DON: Right, but they are not a manufacturer of wind turbines, and they've evolved also from Scottish power. Originally, some of those interests go back to US wind power.

MARC: One of the things I was thinking about is you were at the Department of Energy, and when you left, you wound up developing Combine Hills which is one of the first large-scale wind farms in the eastern part of the gorge.

DON: Well, at the time, there were quite a few other wind projects already up and going in the lower Columbia. The Combine Hills project phases one and two came subsequently in 2004 and 2006, and there were projects well underway across the northwest in the 1990s.

MARC: Okay.

DON: A number of efforts had started even in the 80s. There weren't very many of them, and there were various attempts to permit projects. For instance, on the Oregon coast and different places but on the eastern side of the Cascades, those projects were revisited because the sites were windy, the land was available, and there was the transmission and they were permittable. It hasn't yet proved possible to really permit wind projects on the Oregon or Washington coast.

MARC: Oh, really? So, what is something that you think the general public should know about the development of wind farms and the difficulty that folks encounter doing that if they don't do one-off on their own property? It is because I know of someone that had done that. It wasn't really large, but it was still a good size.

DON: Well, there's a number of aspects where the public needs more education and understanding, and you have to realize that what you're ultimately working with here is human nature. You're working in rural environments where things often have not evolved very fast, and when a wind project wants to come in, it's outside people; it's outside capital. They want to do stuff that if it hasn't been done there before, is new and novel. There's a lot of questions regarding how it's going to look, how it's going to sound, how you're going to continue to farm a ranch around this, and so on. And so, public acceptance and public understanding is a big thing because as an industry, renewable energy is supposed to be sort of the white hat out there. It's environmentally preferable to the fossil fuel and the nuclear industries, and if it's really going to wear that white hat, it's got to not just have the promise of being green power, but it's got to have the practice on the ground, and the practice on the ground is how do you minimize the environmental impact? How do you respect native American values? How do you design the project so that it's aesthetically . . . no one's going to object to it. Well, that might be impossible about "no one," but how do you minimize that?

MARC: Mm-hm.

DON: How do you show people who fear about noise or about pieces of blades flying off and going a mile or things catching on fire all over the place or . . . you know.

MARC: I think we've heard about bird kills and . . .

DON: All of that stuff that gets thrown out . . .

MARC: [inaudible] attitudes.

DON: Sure, and then you have those things because they're our, you know, our front groups and various interests that may like nuclear power or that feel that the jobs at the local coal industry are going to be threatened or coal fire-powered plant will be threatened. Some of these groups get nationally funded on the QT to go out and find examples of poorly done wind farms, of neglect of some of these white hat issues, and mechanical failures that any technology is going to have . . .

MARC: Yeah.

DON: . . . and then exaggerate and generalize this to anywhere and everywhere all the time.

MARC: I remember meeting some folks that you introduced to me some years back whose farm was a recipient you worked with of a couple of these turbine generators, and they seemed really pleased. It seemed that it made their life easier, that it created another revenue stream, and not just based on, you know, cropping and the vagaries of the weather and irrigation, and I wondered if maybe there are some things you can talk about because I've thought this could be something that's a supplemental income to family farms, as I've talked to some people, that can stabilize issues in communities where there are still a lot of family farms that could benefit from this type of expansion of this technology.

DON: Yeah. It's . . .

MARC: It doesn't have a large footprint, per se. You've got a road system, you have the tower, and you have some area around the tower that's protected.

DON: Yeah, what people don't realize is that not only do these wind farms look very elegant in their final form but it also takes an awful lot of work to get to that point, so that's a point of misunderstanding—what all goes into that, how many studies, how much money is at risk, and so on, but the benefits that come back to the landowners and the community don't get as much play as the fears do.

MARC: Mm-hm.

DON: So, if you go, let's say as a developer, to family farms which are really the majority of the farms, even though some of them are corporations or LLCs, we're talking about small entities here where somebody has got a section or several sections of land, and that's about it, and a patchwork of these throughout eastern Oregon and Washington, and for the most part, its dry land wheat. There might be some ranching, scabland, or CRP mixed in here or there.

MARC: CRP. We have people who don't know the anachronisms.

DON: Conservation resource program which is administered by the US Department of Agriculture for preserving natural rangeland.

MARC: Thank you.

DON: And so, instead of farming or ranching it, they plant native plants in there and operate that as a conservation bank, so to speak, and they get paid to do that. So, it's another way that if the land is economically unproductive by normal farming or ranching, it might be eligible for a conservation resource program. Anyway, if a developer goes to a family farmer, the first thing that's going to happen after, of course, is engaging in discussions about how is this going to impact my farming and so on, because that's the livelihood. If that's harmed in any way, they have a vested interest in protecting that and understanding how that's all going to work, but the developer will go and start to pull property records about that property, and so, there are an awful lot of title issues that get identified and cleaned up because if you have title defects for that property, you can't finance placing millions of dollars' worth of wind turbines on it. So, this is one of the things that starts very early and has a lot of title cleanup. Sometimes, that involves surveys. Sometimes, that involves extinguishing old easements. Sometimes, that involves getting things that should've been closed out a long time ago but for neglect or misunderstanding, didn't get closed out, and which are all impairments on the title of that property. So, those things all get scrubbed and taken care of at the developer's expense, and that's permanent for that property whether the developer leaves the next day or they build a project there that sticks. Another thing that happens is that the developer gets into conversations with the landowner, not only about where the turbines are going to be but also about how else we can help this project and what we need to protect here. So, the landowner might have a farm complex or he might have a favorite area down in the ravine where his kids play or there might be a wildlife habitat that the state has got an interest in in this part of it. So, all these things get identified in the project plan, and sometimes, it might be something as simple as oh, we're going to need a zillion tons of gravel and you've got a rock outcropping over there. How would you like it if we open up a rock pit there—a quarry—crush a bunch of rock and so on? And of course, we'll leave it when we're done because we're going to put all this rock on roads for the wind project, and so now a quarry has been developed on this person's property that they get to use themselves, which would've cost quite a bit to set up. You know, often with a little exchange, a big pile of crushed rock gets left there for the landowner to use which he'd normally have to pay a lot for.

MARC: Mm-hm.

DON: And then haul in.

MARC: Right. Ready right there on his property.

DON: Right. So virtually, every wind project has got a rock source or two like that that has fed the property and fed other properties and neighbors around there. I mentioned the road. Well, when the project goes in, a system of roads is put out there that is oftentimes better than the county's gravel roads out on the street. Why? Because that has a pit run, large rock bed, and then crushed rock with binders in it. It's all leveled, highly engineered, and maintained by the wind project, but the landowner gets to use it like it's his own. He doesn't have to pay a dime to have this fabulous road network . . .

MARC: Wow!

DON: . . . created on his property which now he can move his equipment; he can move his wheat haul trucks; he can move fuel. He can move whatever he's doing back and forth on these roads. And what was there before? It was probably not a road at all; it was probably a couple of ruts along the side of a field or on a ridge running through one piece of scab ground to the next.

MARC: That's an enormous benefit to a farmer or landowner . . .

DON: Sure!

MARC: . . . that really stays for some years, maintained at somebody else's expense . . .

DON: Yup!

MARC: . . . all to the benefit, as well, of the farmer . . .

DON: Yup!

MARC: . . . and landowner. Now, did they know this at the beginning that all this can be to their benefit?

DON: Well, this is all part of the conversation that a developer should be having with the landowner upfront, and its part of the project planning, so . . . yeah, they should know what the opportunities are, what, and how things are done. The where and the when are always subject to a subsequent project plan which happens after the surveys and the wind work, and everything else is done. Then, they can figure out exactly where the roads need to go, exactly where the turbines are going to be, and so on. I also want to mention this road network—not only is it maintained, but it's often left for the landowner at the end of the project's life. So, it isn't like cleaned up and we're going to give you back that crappy set of ruts that you were driving in before. No . . .

MARC: Right.

DON: . . . you keep that road network after we're gone. You keep the cattle guards, you keep the gates, you keep the erosion control facilities, the culverts—whatever we've had to put in in association with this road network, those are all . . .

MARC: Permanent improvements.

DON: . . . part of your property's improvements and you're going to use these and we're going to maintain them as long as we're here. And not only is that road network improved inside the project but oftentimes, the developers got to go to the county roadmaster and improve the public roads outside the projects because those roadbeds are not adequate to support the heavyweights of bringing all the wind turbines and cranes and everything across them to build a project.

MARC: Some enormous cranes are used in this business to erect the septic cowling on top of the tower.

DON: Yeah. Enormous cranes and construction activity are very intense. Granted, it looks messy, but it wraps up pretty darn quickly and then everything gets restored according to permit conditions and other conditions, which are in the lease between the landowner and the developer, so everything gets buttoned up again to everybody's satisfaction at the end of the day.

MARC: What then is the result, say at the end of a . . . because it's leased property that the landowner/farmer is receiving from the wind farm—from the developer that there are at least payments for that property.

DON: Right.

MARC: What happens at the end of a lease period and to the turbines? Do they stay on-site? Do they get removed?

DON: Well, there are two ways that that goes. One is the preexisting project when it reaches the end of its either economic, financial, or mechanical life, whichever happens first, then that project is what we call "decommissioned," and the decommissioning means that the project is disassembled and carted away. That will be subject to various requirements in the initial permit that was let for the project in the first place on day one and may also be addressed by provisions in the lease agreement between the landowner and the developer. So, that'll mean that the turbines are taken down and typically would expect that stuff to be hauled off and sold for scrap.

MARC: Mm-hm.

DON: There might be utility improvements, like power lines and what-not. If they're above ground, they'll get taken out. If there's a substation, there'll probably be some discussion with the utility as to where that substation can serve any other public purpose. If it can't, then it'll get taken out. So, we have this decommissioning process. The second thing that might happen is what we call repowering, and repowering basically means the same location you redesign and reconstruct a new project that runs again for another one of these economic, contractual, or financial life cycles.

MARC: Just like upgrades in the technology—take one system that's been running for several years of the property and put another new technology in its place.

DON: Yeah.

MARC: Something like that.

DON: Larger wind turbines with more efficient blades, taller towers . . .

MARC: Yeah.

DON: . . . different control systems, maybe more efficient transformer in the substation . . . who knows? I mean all these things are constantly evolving, and if it makes financial sense to take out this old one and replace it, at some point, somebody will say hey, we need to do this because financially, it's better to have the new stuff in here than to keep running this old stuff day after day. Even though it may be running reliable from an economic point of view, it may be obsolete.

MARC: Is that similar to what's happened in . . . I've seen windmills that look like they'd been running for years in the area of Tehachapi's, in the area down by Palm Springs, and it looks like you have some new facilities that have been installed, but at the same time, you have some much older units still running. Is that . . .

DON: Yeah.

MARC: . . . kind of the scenario?

DON: There are a few places around where you can actually see three generations of wind turbines standing in the same spot.

MARC: Wow, interesting.

DON: One towering above the other, for example—intermixed—and you can see the evolution in the technology right there, but all of these older ones should be eventually decommissioned and the site be restored to something original if there would be no repowering. Now that said, back in the early days, there were a lot of learning curves about what decommissioning really means, what it should entail, and how these projects get maintained and spare parts and broken parts, and so on, over some time. Some permitting authorities were pretty lax about being ahead of the curve on that, and so, some projects are out there even to this day, which did not have those kinds of requirements. They become eyesores locally and become the bad boys that the people who hate wind will go out and find and say, you know, if you do wind power, this is the result, which, of course, if it's the exception rather than the rule and if it applies to a set of permitting regimes and understanding about projects and their regulation that existed in the 80s rather than what would happen today if you tried to do anything today.

MARC: Well, maybe this is, I think, a great summary of how things got started, how you got started in the business as well as some of the evolution that has occurred, and maybe you can help me find a photograph of something you think is an elegant wind farm and we'll get to that to the chapter.

DON: Well, there are plenty of those around the northwest. That's easy to get; just going out highway 84 and 97 and some of the other areas. There are lots of good photo ops.

MARC: Alright, I'll be making that trip.

DON: There is one other thing I wanted to point out about the benefits of these projects too . . .

MARC: Oh, sure, [inaudible] any benefits.

DON: . . . and that is these projects. One of the largest expenses during their running or operations period is the property tax and guess where that goes. That goes into the county.

MARC: That goes to the county coffers.

DON: Yeah. They're the collectors. Now, some portion of it will get siphoned off to the state and portions will get siphoned off to the local school district, to the mosquito abatement, to the police, to the medical services—I mean, you name it—and what's cool about this is that wind projects don't consume public services. You know, when they come in, they don't need any extra schools, they don't need more police, they don't need more hospitals. They don't need all of these things that a lot of other businesses need—let's say a big factory comes in and it brings in a thousand more people to the community—well, now all of these public services go up to serve those thousand people. A wind farm comes in and pays immense property taxes and doesn't need any of those services.

MARC: Now, that's something that I think is largely overlooked.

DON: So, this is pure gravy for the local jurisdictions, and in fact, there are a few instances in the Pacific Northwest where there are enough wind farms in a county that it zeroed out the property taxes for the residents.

MARC: Really? I hadn't heard of that before.

DON: So now, there's so much (laughter)—there's all this revenue coming in. They can afford to do everything including not charge the guy that lives there with property tax anymore . . .

MARC: Wow.

DON: Because the wind project is taking care of this.

MARC: Remarkable. Okay.

DON: So, this is when . . . it's a strong reason why communities ought to be supporting these projects.

MARC: This is a real enormous value in a distributed generation that we've sort of talked about, but we really haven't explained to the public the way they really should grasp these underlying values . . .

DON: Yep.

MARC: . . . to some of the technologies that it doesn't drain resources from the county but it adds considerably to the resources of a county for the community's needs . . .

DON: Yep.

MARC: . . . and education, services, fire, police . . .

DON: Yep.

MARC: . . . libraries . . .

DON: To make some . . .

MARC: I'm glad that I . . . we finally got to talk about this.

DON: Make some examples of that—for example, I know of cases where special EMT training had to be done for high altitude rescue because they're figuring okay, if you've got a wind turbine that has an accident with an employee at the top of that tower, how're you going to get him down? Well, gee, we don't have a local resource and trained people for that. Who do you think pays for that training and the equipment . . .

MARC: The wind . . .

DON: . . . and that local resource? It's the wind companies who've got a project there, and they are now funding those additional capabilities within the county. And, as I say, those public roads, when the wind turbine company improves the public rural roads so that it can haul the equipment in and out, everybody is using those all the time.

MARC: Yeah. So, it's a long-term benefit.

DON: That's right. And in exchange, what do people get? They see a row of wind turbines with some flashing lights at night, and of course, there will be always somebody who doesn't like that—they like it the way it was or it's just not their cup of tea aesthetically, and those will be the people you hear from. It isn't the people who realize all these other benefits. They stay at home and are happy with the benefits.

MARC: Right.

DON: They don't show up at the hearings, they don't write letters to the editor, and so on. You also have this layer of people who will react around the borders of projects. They're afraid of the impacts that'll spill over on their property, like noise or you have human nature. They might operate out of spite because they wanted wind turbines and didn't get any, so I don't want anybody else to be jealous or afraid.

MARC: You mean normal human emotions plugged into this (laughter)?

DON: Exactly. Exactly.

MARC: You can't escape it being human beings and all of that baggage we drag along with us.

DON: Yeah. And so, it's very interesting to see this phenomenon in the context of an industry that has evolved into multi-million dollar wind turbines that are 500 feet tall and with projects that cost $500 million to over a billion dollars that are financed by the biggest banks in the world and fed by giant consulting firms, and so on. Even at that scale, there's still this intensely local process that can't be made a shortcut or scaled like how financing and the technology have been.

MARC: Do you go out in the communities, as we've talked about—you've made those trips to the rural communities and you've looked in the county records and found the problems of title disparities and families that have competing interests as to who's going to get the larger slice of the pie. You've seen quite a few different scenarios, haven't you, from people?

DON: Oh, yeah! And this is why even to this day, most wind projects are still on the scale of just a hundred to a few hundred megawatts. So, even though the technology has scaled and the financing . . . everything has scaled, the size per project, because it's so intensely local like this has not scaled similarly, and you'll see. You know, I mean, there are so many different things that happen. I remember working on one instance where we discovered, going way back, that the USGS town section in range lines in a part of the project area was wrong, and this was published on USGS maps. I mean, it propagated everywhere over the decades, and it was wrong. So, we came in, our surveyor, and we fixed it, and now, this survey problem of the basic land survey that everybody relies on has been fixed; another instance of public roads—we were looking at stringing a power line down a public road, we went out there and started researching how much room we have on the side of the road for the wires and the poles and discovered—guess what? The road that everybody's driving on isn't even on the public road right of way in some places. It's gone off under private land and it's way outside the right of way.

MARC: Really?

DON: And so, you wind up now, once you've identified this, this is a little surprise for everybody. You go back to the county, you work with the county, and they go through a process of rectifying the public road right of way around the as-built road because now, the public has an easement wherever the public road was. So now, that's been fixed. So, you know, you run across these things (laughter).

MARC: When you talk about that problem, it reminds me that there were some folks—I mean because we have lasers and we have GPS today—there was a logging contract in central Oregon where it was having to be on the Kah-Nee-Ta Indian reservation, and it was one of the main reservation boundaries with the forest service on one side and the Indian reservation on the other . . .

DON: Mm-hm.

MARC: . . . and it was discovered that the original survey was off several degrees.

DON: Ouch! Yeah, I remember. It's on the maps.

MARC: And several people don't realize when you go a couple of degrees off of your survey line and you take that out for mile after mile—and I think, in this case, it was almost a hundred miles—that put quite some acres into the Indian reservation and away from the national forest, and it included high-value trees as well, but that got corrected and there was an example. I think that was one of those public stories that we remember that we saw and went, oh my goodness.

DON: Yeah, and that one appears on lots of maps. Here's the original line and here's the corrected line, and you see this wedge that goes for a long way. The other place with a similar problem we had to work with was the wind project—the legal description in the deed describes one of those property boundaries. If you look at the legal description on the property on the other side of the fence, so to speak, well, for that particular boundary, the legal description should match. It should describe the same line. Well, guess what? In this case, it didn't, which created a kind of a no-man's-land between these two deeds for what should've been the same property line. So, we had to figure out who owns that piece of ground.

MARC: Mm.

DON: Is that part of the project leased land or is it part of the neighbor's land? And what brought that to really sharp focus was driving across it because there was a public road that ended at that property line in order to get stuff into the project. Well, where's that property line? It's not the same line. There's this wedge of no man's land. If you can't get access to the site with your equipment, how can you build a project?

MARC: Yeah.

DON: So, a simple thing like this can snowball into big problems, and they get identified through, you know, a careful review of the records to see that everything is correct.

MARC: Yeah.

DON: So, in that instance, you know, we wound up having to get lawyers involved for both property owners. They, in turn, had to choose a surveyor who also had to choose a third surveyor to come in and say where that line truly was, and then, we all had to work back through lease revisions and property description revisions and get that all recorded, and then, finally now, the deeds on both sides of that line said the same line.

MARC: Well, that sounds like a bit of a night . . . a huge headache and a nightmare.

DON: Yeah, and these things take months to resolve, and they're total surprises.

MARC: You see, that's the other part about this whole business in energy. It takes years sometimes . . .

DON: Yeah.

MARC: . . . for these projects to materialize from. Let's try and put a wind farm over there, and it's going to be years if it even gets completed.

DON: Right, and that's where there's a lot of what I'll call project failure. It's not that the turbines don't work or aren't economic. It's rare that there isn't enough wind because they wouldn't have started there if there wasn't in the first place—it's usually because of what we call these "soft factors." You find an environmental issue, you find a legal issue, you find a community acceptance issue, and you find a permitting issue. These are the major reasons for project failures, yet as an industry, we're all focused on what's the newest turbine blade or what's the newest best gearbox and what's all of the tech parts of it, but really, where we need to improve is over in the soft part.

MARC: Interesting.

DON: Because that's what makes projects—that's what the sources of project failure normally are. It's on that side. So that's a place where a lot has to be done. And speaking of all these kinds of issues that you're dealing with in the interim, as a developer, you need to identify these and resolve them as quickly and as early as possible because every day that you're there trying to make this happen, money is being spent and it's all at risk, and the very worst nightmare is that you neglected something—you've got your permits, you've got your power cell contract, you've equipment ordered, and now you're going through term financing, you've got a roomful of lawyers that's costing everybody many thousands of dollars an hour, and they're doing due diligence on every little aspect of the project to make sure that every risk is identified, and for every identified risk, they're all looking around the table. Who's going to eat that risk? Who's going to take it? And so the last thing you need is from some surprise coming up because some due diligence consultant went out there and found that, oh, by the way, where you're trying to put your power poles isn't where you thought it was or you can't get across that piece of ground there because that's no man's land and you don't have an agreement or a right to do that, or one of these other kinds of sneaky little issues. And so at that point, this conversation stops in the room, and they say okay, here's this problem, and we're not getting any further with this financing until either (a) it's fixed or (b) somebody around the table eats the risk, which means that if that kills the project, then you're the one who's going to pay everybody else for their costs and wasted money that has been spent so far. Somebody's doing to eat that risk; they're going to take it. They're going to take liability for this problem that wasn't fixed. So, that's the worst nightmare scenario, and this is why when you're doing careful development, it's an exercise in risk management from day one. You're looking for all of those things, you're trying to identify them, trying to understand them, trying to get rid of them, so this is another hidden job that people don't realize and understand about the projects. It's not just straightforward, let's go build it. It's a risk management exercise. And on the flip side of that same coin, you're creating an asset, so sometimes, people ask me what business am I in. I can say, "I'm in the asset-creation business" just like I can say, "I'm a project manager."

MARC: A risk assessment.

DON: Or some of these other things.

MARC: Risk analyst.

DON: You're wearing a lot of hats when you're doing this, basically.

MARC: Yeah, I recognize that. I had a few of those experiences myself.

DON: Yeah.

MARC: Well, Don, I just, you know, don't want to . . . I just think this is a great introduction, and you've explained the wind scenarios in an in-depth way that I knew you would and far better than I could've ever commented on. I really appreciate your coming up here and sitting down with me to add this piece to my work on renewable energy.

DON: Well, I'm happy to contribute to that work because, you know, these are things that all need to be told and people need to understand this if we're going to be more successful at all of this business.

MARC: Well, we have the technology, and as you're pointing out, it's the soft issues that really compound the complexity, and if people aren't beginning to get educated on, hey, we're trying to do the right thing for everybody in a community that has real benefits in our renewables—the value of our renewables—it'll be easier as we have this opportunity to explain to people and pretty direct language from our experiences because there aren't a lot of people with . . . now, what're we saying . . . from the early 70s for us. You know, you and I have, well, what's that going to be? Forties—let's see—30, 45 . . .

DON: Forty-seven years.

MARC: Combined between us, it's something like 90 years' experience?

DON: Yeah. Yeah, it's mostly a lot of young people who are going to learn a lot of this stuff the hard way, and it's not necessary to learn it the hard way, and you can't get taught this stuff in engineering school. They're not concerned with this.

MARC: Right. Yeah. Yeah. Well, once again, Don, thank you. This was with Don Bain of Aeropower in Portland, Oregon, and this is Marc Rappaport working on my book for renewable energy.

DON: Good luck with your book.

MARC: Thanks, Don.

Made in United States
Troutdale, OR
11/03/2026

56845371R10069